René Carré, Pierre Divenyi, Mohamad Mrayati

Speech: A Dynamic Process

Also of Interest

S. Weber Russell, 2015
Computer Interpretation of Metaphoric Phrases
ISBN 978-1-5015-1065-6, e-ISBN 978-1-5015-0217-0,
e-ISBN (EPUB) 978-1-5015-0219-4, Set-ISBN 978-1-5015-0218-7

B. Sharp, R. Delmonte (Eds.), 2015
Natural Language Processing and Cognitive Science: Proceedings 2014
ISBN 978-1-5015-1042-7, e-ISBN 978-1-5015-0128-9,
e-ISBN (EPUB) 978-1-5015-0131-9, Set-ISBN 978-1-5015-0129-6

J. Markowitz (Ed.), 2014
Robots that Talk and Listen: Technology and Social Impact
ISBN 978-1-61451-603-3, e-ISBN (PDF) 978-1-61451-440-4,
e-ISBN (EPUB) 978-1-61451-915-7, Set-ISBN 978-1-61451-441-1

A. Neustein (Series Ed.)
Speech Technology and Text Mining in Medicine and Healthcare
ISSN 2329-5198, e-ISSN 2329-5201

René Carré, Pierre Divenyi, Mohamad Mrayati

Speech: A Dynamic Process

DE GRUYTER

Authors
René Carré
Université Louis Lumière
Laboratoire Dynamique du Langage
Lyon, France
recarre@orange.fr

Pierre Divenyi
Stanford University
Center for Computer Research in Music and Acoustics
Stanford, California 94305
USA
pdivenyi@ccrma.stanford.edu

Mohamad Mrayati
United Nations Development Program
Diplomatic Quarter, Riyadh
Saudi Arabia
mmrayati@gmail.com

ISBN 978-1-5015-1060-1
e-ISBN (PDF) 978-1-5015-0201-9
e-ISBN (EPUB) 978-1-5015-0205-7
Set-ISBN 978-1-5015-0202-6

Library of Congress Cataloging-in-Publication Data
A CIP catalog record for this book has been applied for at the Library of Congress.

Bibliographic information published by the Deutsche Nationalbibliothek
The Deutsche Nationalbibliothek lists this publication in the Deutsche Nationalbibliografie; detailed bibliographic data are available on the Internet at http://dnb.dnb.de.

Cover image: threebit / iStock / thinkstock
Typesetting: le-tex publishing services GmbH, Leipzig
Printing and binding: CPI books GmbH, Leck
♾ Printed on acid-free paper
Printed in Germany

www.degruyter.com

Preface

Speech is an activity of overarching importance, a natural tool that essentially makes communication between individuals possible. Because of its weight but also because of its high complexity, to gain a thorough understanding of speech has intrigued scientists, philosophers, and thinkers for centuries. Its intricate nature means its study needs to be undertaken within and across several disciplines – physics, linguistics, motor and sensory physiology and psychology, mathematics, information theory, communication theory, social studies, ethology, evolutionary history and theory, just to name the top contenders. This broad interest in speech has translated into over five centuries' worth of research published in books, treatises, scientific journals, and conference proceedings. In addition, several scientists also undertook the task of presenting their overall, explanatory view of speech. Because both the scope and the angle of approach of these publications were often quite different, and also divergent, their range is fairly broad. Some of them were written mainly for the classroom; they mostly expose the overall process and the basic mechanisms of speech production and/or perception. Other writings, among them several highly respected publications, present an in-depth view of speech through a particular lens, with the focus being on process, mechanism, mathematical formulation, theory, or some other aspect of concern. Although many of these were intended mainly for the specialist or an educated audience, anybody seriously interested in the topic could read them to learn, to obtain specific information, or to simply become inspired.

A quick survey of the literature on speech indicates that the list is fairly long: in the in-depth category alone, the volumes would easily fill a small bookcase. So, one is justified to ask the question: given all this literature – dealing with speech, whether parts of it or its entirety – can there be anything new to offer? However, we feel that there was a compelling reason for us to write the present volume aimed at taking a rarely trodden path to address a single intriguing question: based only on physical principles and the aim of obtaining acoustic changes at minimum energy cost, can the use of a simple tube evolve, step-by-step, into an optimal acoustic production system comparable to the one we use when we speak? Can the sounds obtained by such a production system be perceived as speech sounds? Can this approach lead to an "acoustic phonology" corresponding to vowels and consonants?

In our work we replace the vocal tract with a simple, flexible, uniform tube of the same length but forget everything else about the human apparatus, set the clock to an artificial "time zero," and allow deformation of the tube to be powered solely by the above guiding forces. Then, the primitive process of production

DOI 10.1515/9781501502019-202

efficiently evolves into an optimal, dynamic system. By dynamics we mean that it is the acoustic changes that constitute the information; by efficient we mean that, to generate the largest possible acoustic variation, such changes are produced by exerting minimum effort on deforming the tube; by optimal we mean that modeling, control and use of the tube are implemented in a simple manner that achieves maximum efficiency in the production of the most important classes of speech sounds.

These are the arguments that are opened up, discussed, and explained throughout the book. Going over them chapter by chapter will be a journey that the reader may find informational and even original but also, we hope, ultimately interesting and pleasant.

Contents

To Gunnar Fant
scientist, teacher, friend

Introduction

The study of any scientific topic can adopt one of two main lines of reasoning. The most often used one, *inductive* reasoning, can be described by the sequence (1) initial data collection, (2) incremental, further data collection to explore the effect of different parameters, settings, etc., (3) building a mathematical model to explain the data, (4) using inferential statistics (Cox, 1961) on the observed and the modeled data to identify trends and obtain predictions, (5) modifying the model to account for discrepancies between prediction and data, and so on. This reasoning builds science from the bottom up (i.e., from the most basic to increasingly more complex levels) – a process not without merits because the data it creates picture reality. However, inductive reasoning has many shortcomings (for a list see, e.g., Vickers, 2014). For one, the predictions it generates are fraught with inherent circularity: regularities in one specific set of data are used to predict regularities in other sets of data. Moreover, it is unable to yield strong explanatory conclusions because it is not designed to identify causal and generally simple basic principles or laws that underlie the observed facts. Furthermore, it can model the data only in the statistical sense, which becomes increasingly unreliable as the complexity of the phenomenon under study increases. To account for these complexities, if such models are to succeed[1], they tend to become more and more convoluted, in direct opposition to the principal characteristic of scientific theories – simplicity ("Occam's Razor," Soklakov, 2002) – and to the primary *raison d'être* of scientific models – finding the simplest way to capture the essential in the data (Rasmussen and Ghahramani, 2001). All this is not to say that inductive reasoning has not produced worthwhile projects. In the field of speech, construction of databases – speech sound inventories across many languages, complete with classification and statistics – exemplify extremely useful work accomplished with this approach. For instance, Peterson and Barney (1952) have provided a still much used spectral representation of American English vowels spoken by a large number of individuals and carefully measured the first two resonant frequencies (the first two formants, F_1 and F_2) of each vowel. Their data-driven approach was able to show that, despite a large variability in the data, vowels illustrated in the F_1-F_2 plane outline a triangle; yet, it could not answer fundamental questions like how the vowel triangle came into existence.

The *deductive* line of reasoning, follows a different sequence: (1) Observe a particular existing construct with all its multiple, complex features, (2) find a sim-

1 A comment John Ohala attributes to James Kirby, made after Kirby presented his population model of language change (Kirby and Sonderegger, 2013).

DOI 10.1515/9781501502019-001

ple universal principle, law, or theory that, when applied to the construct, would replicate it, (3) demonstrate, by applying logical reasoning, that the construct actually derives from that principle, law, or theory, (4) translate that flow of logic into a simple formal model that will be able to generate universal predictions because the principle or law on which it is based is universal, (5) verify if the model's predictions are in agreement with a comprehensive panoply of classic and new data. This line of reasoning is deductive and *explicative*: it builds a scientific argument from the top down and then returns to the top again. Predictions generated using this approach are able to identify causal principles underlying observed results. Deductive reasoning is especially well suited for helping to understand how an observed phenomenon can be logically explained by way of a dynamic process. This approach makes it possible to ask questions such as how the phenomenon arose in the first place and how it may have developed over time, whether other related phenomena can be predicted using the same logic and what underlying principles guided its evolution. On the whole, starting from general laws or principles leads to simple explanations of observed phenomena, and simple explanations often become piers of complex edifices. Quoting Jean Perrin, 1926 Nobel Prize of physics, "Science replaces the complicated visible by the simple invisible." (See Figure 0)

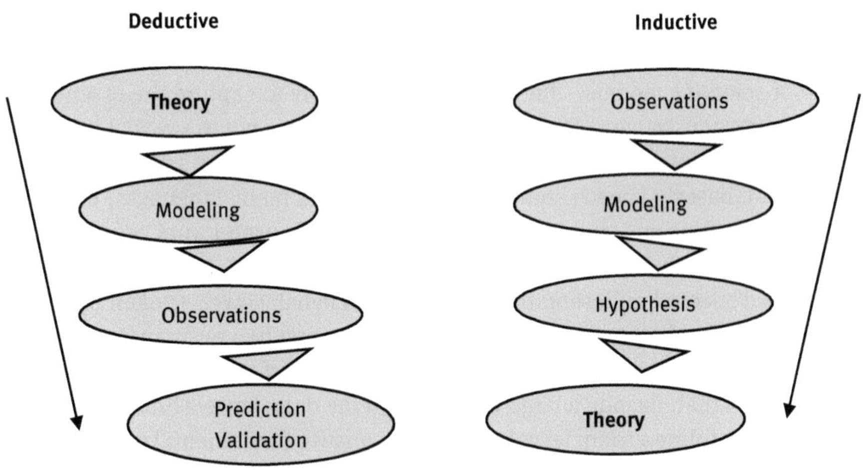

Fig. 0: Schematic representation of deductive and inductive reasoning.

In this volume we use deductive reasoning to put the human activity of speech under a magnifying glass. To understand how human beings communicate using the complex acoustic waves that speech represents, it is obviously necessary to examine the acoustics of the speech signal as well as the human systems that produce and perceive it. But, once arriving at a coherent description of speech in terms of its physical and physiological attributes, more fundamental questions arise: By virtue of what cause and for what purpose do the biological systems of production and perception of speech possess their particular characteristics? Are these systems optimal for satisfying speech communication needs? To find an answer to these questions, we want to take some elementary mechanism guided by some general, higher order rules, and by showing that, driven by those principles applied to this mechanism, a universal and efficient acoustic system will arise. As for the mechanism, we chose a sound-producing 18-cm-long flexible acoustic tube – an analog of the male vocal tract[2] – and we adopted two general rules: the information theoretical principle of achieving maximally efficient communication while expanding minimum effort (i.e., highest entropy increase at the cost of minimum energy) and the principle of simplicity (small number of parameters to define the mechanism and use of the simplest code to represent information). Starting from an initial uniform cross-sectional, neutral, tube shape and relying on the tube's natural resonant frequencies, we will have the system evolve step-by-step through an iterative process without any additional constraint and arrive at an efficient and optimal acoustic production system. Each step of the process is efficient, that is, it consists of minimum shape deformation leading to maximum acoustic variation. The most important property of the process is that it is inherently and fully dynamic: the information in the sounds it generates resides in the changes of the acoustics rather than in a chain of static, discrete events. In such a framework, information dwells in the properties of the ongoing deformations of the tube's shape or, equivalently, in the properties of the continuous changes of the tube's resonances. Any specific acoustic change is obtained by applying a sparse code that deforms the tube in specific ways: the code can portray a given formant transition by specifying its point of departure, its direction, its velocity, and its duration. *Dynamics* – the constantly changing flow of information – constitutes the most essential feature of the production process and of the system we present in the book.

In summary, this volume presents an efficient dynamic production system deductively derived from physical characteristics of a tube having the length of the male vocal tract. The *production* system is optimal for generating efficient

2 At the origin this system may not have been well adapted for acoustic communication, in view of the fact that the same tube was, and still is, used for a variety of other tasks, such as drinking, breathing, etc., but primarily for efficiently eating. Also, although we chose to work with the male vocal tract length, results of our computational system apply to women's and children's speech production with few scaling modifications.

acoustic trajectories. The maximum acoustic space of the tube naturally derives from its dynamics, although it also allows forming sub-spaces to guarantee *sufficient* acoustic contrast adapted for communicating under diverse acoustic environments. When it is paired with a perceptual system ideally suited to decoding dynamic information, the production system, the tube, represents an optimal *communication* device. The book's principal purpose is to show and illustrate that the sheer existence of the tube and its use as a dynamic production device create a physical system comparable to the biological production system that we use when we speak. It demonstrates that the creation of two categories of sounds we have explored in detail – oral vowels and plosive consonants – is an inherent consequence of the tube's properties. The dynamic approach we have taken predicts and explains the production and perception of these two sound categories, although extension to other speech sounds is also discussed.

To keep the book's focus on these matters, there are some areas into which we could not delve, despite our recognition of their importance. The omitted topics include features like nasality, laterality, and fricatives; source, especially the vocal folds but also frication noise; prosody and the speech of the singing voice; diachronic phonological sound change and evolution. While extending our system to incorporate these topics would have been of great interest to us, they would need to be treated in a future volume, first because of the lack of a number of required underlying complementary studies, but mainly because we wanted to keep this book focused on our approach and the consequences that ensue from it.

The book follows the above-described process in ten chapters, unwinding and validating it stage by stage: from basic principles to the physical reality of the tube, to tube deformation properties and data, to a model generating similar data, to testing the model by comparing its output to results on speech production and perception, to revealing it as a dynamic system, to viewing its output in the perspective of the auditory system, all the way to discussing its epistemological implications, proposed extensions, and possible applications.

Chapter 1 (**Speech: Results, theories, models**) unwinds the chain of ideas, observations, and reflections that has led to the book. A brief survey of the main theories, models, and findings of speech – in the articulatory, acoustic, phonetic, and perceptual domains covered in this book – are presented to serve as a comparison to our approach. To make it accessible also to the non-speech-specialist reader, we tried to expose this chain in sufficiently general terms.

Chapter 2 **(Perturbation and sensitivity functions)** explains what these functions are and how they are used in the following chapters. In a deductive approach, *sensitivity analysis* offers a powerful tool for investigating a complex system. In this chapter, we recall the way these functions are derived and used, specifically for the acoustic tube.

In Chapter 3 (**Efficient acoustic production system**) we derive and explain the characteristics of the 18-cm long tube used for the generation of the acoustic signals. The characteristics are obtained from an iterative process consisting of step-by-step deformations of the tube following the simple rule that each small deformation must result in maximum acoustic variation at the cost of minimum effort or, equivalently, a minimum deformation provokes a maximum acoustic variation. Guided by this approach, distinctive regions of the tube and synergetic deformations are obtained. Because this process is based on deformations in constant motion rather than on a succession of static targets, it is intrinsically dynamic.

Chapter 4 (**The Distinctive Region Model**) describes a model developed using the results of the deductive process obtained in Chapter 3. Properties of the model satisfy all requirements for building an efficient acoustic production system, such as distinctive regions, efficient shape deformations, and synergy. The chapter also formally demonstrates the existence of privileged acoustic trajectories that can be used as coding primitives of the system. The model actually acts as a formal link between the tube deformation processes outlined in the previous chapter and the speech production processes presented in the following chapters.

Chapter 5 (**Speech production and the model**) compares characteristics of speech production and those of the model. The chapter demonstrates that data generated by the model are in remarkable agreement with speech production results: both display maximum formant spaces outlining the same vowel triangle, and both show identical gestural deformations, identical synergetic deformation of the tube, and identical places of articulation for vowel and consonant production. The good agreement implies that oral vowels and plosive consonants are intrinsic physical properties of the acoustic tube and, therefore, that those properties must have existed as such before languages appeared. This chapter takes the formant transitions and the acoustic trajectory primitives produced by the model, places them in a syllabic frame, and generates vowels as dynamic trajectory states rather than static targets. It also provides a deductive explanation of human speech production and presents it as the consequence of specific, efficient deformations of the tube performed with the aim of achieving maximum acoustic variations.

Chapter 6 (**Vowel systems as predicted by the model**) shows that vocalic trajectories obtained by the model predict vowel systems observed in a large number of languages. The actual vowel system in a given language depends on the number of vowels it uses, though some vowel inventories may be decided by sociolinguistic factors. The agreeement between the actual and the predicted vowel systems can be explained by the model that shapes the acoustic plane into a limited number of crisscrossing vocalic trajectories. The actual vowels in the inventory of a given language always fall on these trajectories.

Chapter 7 (**Speech dynamics and the model**) focuses on the inherently dynamic nature of speech such that information emerges from transitions rather than from successive static states. This dynamic representation confers an overriding importance to the dimension of time, movement, synchrony between primitive gestures, the primary role of which is illustrated in examples of vowel transitions and syllabic co-production. Experiments using pseudo-vocalic trajectories outside the vowel triangle clearly demonstrate that the slope and the duration of the transition are parameters used by the perceptual system to identify, categorize, and discriminate vowels.

Chapter 8 (**Speech perception viewed from the model**) first describes principal static and dynamic characteristics of the auditory system and makes the case that it is optimized for speech perception. Then it describes properties of the perception of both amplitude and frequency modulations before an exposé on auditory patterns and the perceptual consequences of the simultaneous presence of more than a single sound: energetic and informational masking and, especially, auditory scene analysis. This part segues into another on a topic laying the auditory foundations of the perception of trajectory dynamics in speech: the perception of monotonic frequency transitions. Finally, theories and results on the perception of temporal segmentation are presented and discussed in their relationship to speech perception. The chapter argues for links between perception and production of speech in ways consistent with the model's predictions.

Chapter 9 (**Epistemological considerations**) discusses epistemological implications of the general deductive, iterative, and modeling approach and results, as laid out in the previous chapters. It also brings up some relevant theoretical considerations, discussing their relation to recent findings and trends in phonology and phonetics.

Chapter 10 (**Perspectives and conclusions**) summarizes the main findings and conclusions of the book and outlines extensions of the findings, the methodology, and the model to different aspects of speech, such as phonology, as well as speech sounds not discussed in the previous chapters. It also brings up further issues on temporal normalization, deconvolution of co-produced gestures, and human speech evolution. It then looks forward and discusses possible applications.

Finally, we wish to acknowledge the contribution of all those who gave us support and help in our effort to put this volume together. First, writing it, even thinking about writing it, could not have been possible without the three of us climbing, with great respect, on the shoulders of the scientists who preceded us and who had asked similar questions. We were especially inspired by the seminal work of scientists such as Valery Kozhevnikov, Ludmilla Chistovich, Gunnar Fant, James Flanagan, Björn Lindblom, Manfred Schroeder, and Kenneth Stevens. We have also greatly benefited from extended interaction with our colleagues Alvin Liberman, Michael Studdert-Kennedy, John Ohala, and Carol Fowler. We are especially thankful for the guidance we have received from Max Wajskop and Mario Rossi, and for the numerous illuminating discussions we were fortunate to have had with our colleagues and co-workers Jean-Marie Hombert, François Pellegrino, Jean-Sylvain Liénard, Jean-Pierre Tubach, Shinji Maeda, Didier Demolin, Egidio Marsico, Eric Castelli, Gérard Chollet, and Steven Greenberg. Also, we want to thank Michael Studdert-Kennedy and John Ohala for their critical reading of an earlier version of the manuscript, as well as Leonardo Milla, Lena Ebert, George Powell, and other members of the devoted editorial team at de Gruyter for providing scrupulous, efficient, and personable assistance throughout the publication process. Finally, we would like to express our gratitude for the patient and tolerant support of our respective partners during our frequent physical and mental absences while working on the book.

1 Speech: results, theories, models

1.1 Background

Speech is a communication process: a talker-to-listener flow of information traversing an often noisy media. From an informational viewpoint, this flow has to be a sequence of events that are choices between alternatives. Of course, as Claude Shannon (1948) demonstrated in his classic but forever young and fresh paper on information theory, the flow is not forced to consist of Morse code-like binary or alphabet-like multi-valued discrete elements; it may be an ensemble of features describing continuous processes, such as those in a noisy telephone conversation. Interestingly, the linguist Roman Jakobson was among the first to understand the value of Shannon's system as he re-formulated his original treatise on distinctive features of phonemes (Jakobson, Fant, and Halle 1951) to correspond to information-theoretic concepts (Jakobson and Halle, 1956), but it was Alvin Liberman and his colleagues who first talked about the "speech code" (Liberman et al., 1967). Yet, it should be clear that the above flow constitutes only one of the two arches of communication, the other being the explicit or tacit indication by the listener to the talker that the information was received and decoded correctly. In the dynamic framework, information is defined, among others, by properties of gestural deformations of the vocal tract's area function or, equivalently, by the continuous change of formant resonances. We can apply a sparse code for communicating articulatory motions via the resulting acoustic changes: the code can portray a formant transition by specifying its point of departure (most often a time-domain discontinuity of a formant trajectory), its direction, its velocity, and its duration, and send a packet of bits only when any of these properties changes. Although such a communication framework does share features with several of the computational speech analysis systems currently used for automatic speech recognition (ASR, e.g., Deng, 2006), and although those systems do represent and code information in the form of spectral changes in the speech waveform, the system we will present is significantly sparser. When it comes to information, we also believe that production had to evolve using decision trees, which first had to be simply binary and which then had to expand logarithmically by factors of 2, as Shannon's (1948) original system predicts.

But, as we have already said in the Preface, it is *dynamics* – the constantly changing flow of information – that constitutes the most essential feature of the communication process and of the system we are about to describe. Our interest in the dynamics of speech production and perception was piqued early on when we became exposed to the off-the-beaten-track studies by Ludmilla Chis-

DOI 10.1515/9781501502019-002

tovich and her collaborator-husband Valery Kozhevnikov (Kozhevnikov and Chistovich, 1965; Chistovich et al., 1972). This couple, while investigating the known phenomenon of gesture anticipation in coarticulated consonants, was the first to suggest that speech should be studied as a dynamic flow of syllables rather than as a sequence of phonemes. Under the Soviet system much of this research was classified and to this date only a small fragment of the vast productivity at the Pavlov Institute's speech laboratory has been translated from the Russian.[1] Sadly, as first-rate scientists as Kozhevnikov and Chistovich were, they are still largely unfamiliar to Western audiences, and their work has remained outside the main body of current scientific knowledge.

Above all, we are greatly indebted to Gunnar Fant from whom we learned so much about speech acoustics and production, by way of reading his extensive work and also through invaluable personal contact. His seminal book (Fant, 1960) has been a constant source of reference while writing this volume. We also owe much to the work and ideas of Kenneth Stevens and James Flanagan, two scientists whose contributions to speech science are universally recognized. As anybody who had any interest in learning speech acoustics beyond the basics, we were students of their respective treatises in this field, "Acoustic Phonetics" by Stevens (2000) and "Speech Analysis Synthesis and Perception" by Flanagan (1972). Our book has also benefited from Stevens' quantal theory of speech (Stevens, 1972) that defined articulatory-acoustic relations in dynamic terms. The model we shall describe in Chapter 4 rests on foundations laid by Manfred Schroeder, a physicist whose research had a significant impact in several branches of acoustics. We particularly appreciate his elegant mathematical treatment of area function dynamics and his solution (one of the earliest) to the thorny problem of converting articulatory motion to acoustics (Schroeder, 1967; Schroeder and Strube, 1979). We gladly acknowledge Sven Öhman's contribution to our view on the dynamics of coarticulated consonant–vowel (CV) and vowel–consonant (VC) utterances (Öhman, 1966a, 1966b), from the points of view of both production and perception. Finally, we have learned much from Hiroya Fujisaki's writings about his imaginative experiments and precise models on the perception of static and dynamic speech sounds (Fujisaki and Kawashima, 1971; Fujisaki and Sekimoto, 1975; Fujisaki, 1979). Our understanding of the importance of articulatory dynamics was also stimulated by Björn Lindblom's "H & H" (hyper- and hypo-speech) theory (Lindblom, 1990b) and vowel system prediction

1 Much of that research was published over a 20-year period starting around 1965 in Russian but only a small portion of it was translated into English, with some of the rest available from secondary sources.

(Lindblom, 1986b) – a linguist of broad interests with whom all three of us had the good fortune to interact over several years. We also credit him for his investigations on the prediction of vowel systems, and on incomplete vowel transitions ("vowel reduction," Lindblom and Studdert-Kennedy, 1967). Incidentally, it was Lindblom who first championed taking a deductive approach for the study of speech (Lindblom, 1992). We also benefited from the pioneering research of investigators at Haskins Laboratories – Alvin Liberman, Frank Cooper, and many of their colleagues – who, using their innovative spectrogram-to-audio "Pattern Playback" talking machine, were among the first to study the role of CV and VC transitions in speech perception (Delattre et al., 1955). Although still focused on the essentially static phoneme as the smallest element of speech, as stated by the motor theory, Liberman (1957; 1967) nonetheless viewed speech as a dynamic process, and so did Fowler (1986) when stating the dynamic property of perceptual direct realism. However, it was Catherine Browman's and Louis Goldstein's hypothesis (1986) positing a direct relation between articulatory movements and phonological representation (articulatory phonology) that moved the motor theory to a new level at which dynamics became essential – as it is for our own approach. Articulatory phonology is an attempt to represent the speech code in terms of articulatory gestures. Associated with a motor equivalence mechanism (Saltzman, 1986; Guenther, 1995), this approach allows the study of respective roles these gestures play in diverse speech phenomena, such as coarticulation, vowel reduction, etc. This approach was also encouraged by the then-surprising results of silent consonant experiments by Winifred Strange and James Jenkins (Strange, Verbrugge, Shankweiller, et al., 1976; Strange et al., 1983). Their experimental results, confirmed by automatic speech recognition tests by Furui (1986b), showed that transitions in nonsense CVC (consonant-vowel-consonant) syllables are sufficient for the recognition of a silenced vowel or silenced initial and final consonants (see also Hermansky, 2011).

In recent years the view of speech as a dynamic process has gained an ever-increasing number of adepts and many studies have demonstrated that, when listening to speech, people recognize the dynamic progression of sounds rather than a sequence of static phonemes. This dynamic view has been given support by neurophysiological results showing that perceptual systems respond principally to changes in the environment (Barlow, 1972; Fontanini and Katz, 2008) and has become a cornerstone of kinetic theories of speech perception (Kluender et al., 2003). At the same time, more precise measurements have made it possible to discover vowel-inherent spectral changes (e.g., Nearey and Assmann, 1986) in vowel formant frequencies once considered static (Morrison and Assmann, 2013). Interestingly, while formant frequencies measured at the midpoint of natural American English vowels display large inter-speaker variability (Peterson and Barney,

1952), CV and VC formant transitions for a given vowel and a given consonant are remarkably stable across talkers (Hillenbrand et al., 1995, 2001).

Nevertheless, the human acoustic *communication* system could not be optimally efficient without the articulatory apparatus being paired with an efficient receiver – an auditory counterpart capable of detecting and discriminating both the sounds produced and the dynamic flow of the articulatory information. But we know that the vertebrate auditory system, including the human one, had developed into the fine device it is – probably a lot earlier than the human vocal tract evolved to serve the purposes of speech communication – mostly to detect and identify predator, prey, and sexual mate. So, the two pertinent questions are: (1) What particular auditory capabilities are needed for the detection, discrimination, and identification of sounds emitted by a possibly distant talker communicating possibly in an environment where the speech information is often masked by noise? And (2) are the particular auditory functions that comprise those capabilities fine-tuned to perform optimally when the parameters of the functions match those of the sounds produced dynamically by the articulatory apparatus?

To answer these two questions, first one should list the speech signal's physical characteristics and describe them from the listener's point of view. A good point of departure is Rainer Plomp's characterization of speech as being a signal slowly varying in amplitude and frequency (Plomp, 1983). In other words, speech is a modulated signal and, in order to perceive it, the auditory system needs more than effectively dealing only with basic auditory attributes – such as audibility, signal-to-noise ratio (SNR) at threshold, intensity, frequency and pitch, timbre, duration, masking, localization – a list that about 150 years of research in psychoacoustics has been attacking with greater and greater success. On the whole, it seems that the human auditory system is an excellent match for analyzing the speech waveform: the normal speaking intensity at conversation range is sufficient to transmit information even for a whisper, frequency resolution is finer than the one necessary to resolve neighboring formants for all vowels and consonants, pitch resolution is so exquisite that minute prosody-driven fundamental frequency fluctuations (signaling some subtle emotional change) can be perfectly decoded, gaps of only 1–2 milliseconds in duration can be recognized, horizontal localization of a source is good enough to track even slight displacements of a talker. But, thinking of Plomp's definition of speech as a modulated waveform, it is necessary to investigate the perception of both amplitude (AM) and frequency (FM) modulations: AM for understanding the mainly syllabic-rate fluctuations of the speech waveform, and FM specially to study the perception of formant frequency modulations (i.e., sweeps). The earliest measurements of dynamic changes in the formant frequencies of natural speech are credited to Lehiste and Peterson (1961), although perceptual effects of formant frequency changes in both natural and synthesized vowels were also investigated early on by the Leningrad group (Kozhevnikov and Chistovich, 1965; Chistovich, 1971; Chistovich, Lublinskaja, et al., 1982). When presented with non-speech filtered harmonic series single-formant (vowel-like) tone complexes, trained listeners are able to perceive very rapid formant frequency changes at very short, and low-velocity changes at a little longer but still quite short, durations (e.g., Schouten, 1985). These experimental results shed light on how we must be perceiving speech: for natural speech samples formant changes are inherent even in vowels that, for a long time, have been considered steady-state (Nearey, 1989).

In real life settings, however, we are often faced with the task of listening to a specific talker when in the background other persons are also talking (a situation named the "cocktail party effect" by Colin Cherry [1953; 1966]). Interference in the perception of one signal by another has been studied for a very long time; first, it was referred to as a higher intensity interferer, mostly noise, masking the audibility of a lower intensity signal. Such masking, however, was more recently identified as only one of two kinds of interference, the other being one in which the masker does not abolish or diminish detectability of the signal but decreases its identifiability or discriminability – i.e., it lowers the ability to extract information from it. Appropriately, the first kind is now called "energetic" and the second "informational" masking (see Durlach et al., 2003, for a thorough explanation of the difference between the two). But masking in cocktail party situations is more complicated: there we have the AM/FM "noise" of the talkers in the background interfering with the AM/FM signal of the target talker's speech. Mutual interference of complex AM or FM has been examined using both monaural and spatial presentation methods by several investigators in the recent past (e.g., Yost et al., 1989; de Cheveigné, 1993; Kidd et al., 1994, 2010; Moore et al., 1995; Lyzenga and Carlyon, 1999; Divenyi, 2005b), in order to explain the nuts-and-bolts of the cocktail party problem. Trying to solve that problem has high importance in view of the fact that hearing loss and aging (the latter even without abnormal hearing thresholds) both lead to often severe impairment in the understanding of speech in a cocktail party setting. Our work in the area (Divenyi and Haupt, 1997c, 1997b, 1997a; Divenyi, 2005b) has also demonstrated that age-related loss of cocktail party performance is related to loss of the ability to follow, in the midst of similar interference, fluctuations of both AM (Divenyi and Brandmeyer, 2003) and FM (Divenyi, 2005b, 2014) signals.

In the remainder of this chapter, results, theories, and models most relevant to establishing the bridge between acoustic, phonetic, articulatory, and perceptual areas of speech are presented, as illustrated by the cartoon in Figure 1.1 (see Carré and Mrayati, 1990, for an overview). To keep this retrospective of manageable extent, it is focused on how work in these domains has helped create an understanding of the system of speech communication that we use in our daily life.

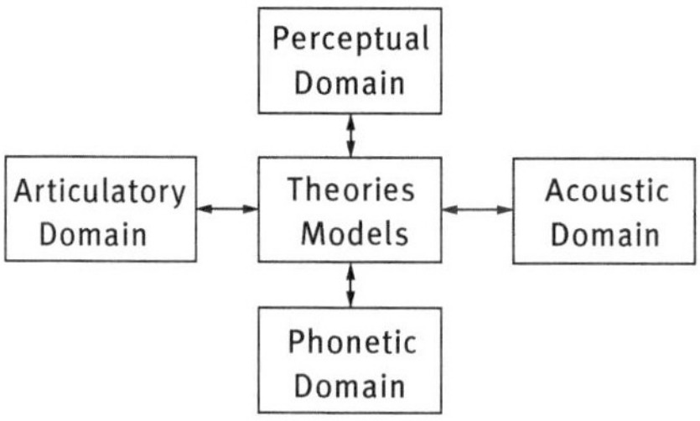

Fig. 1.1: Schematic representation of speech domains.

1.2 Speech production

Although the remainder of this chapter may resemble a literature review, in fact, it only attempts to reference seminal research that we consider representative in terms of the basic underpinnings of our approach, without aiming at an exhaustive coverage of even those areas that we treat. The objective of the following pages is to highlight previous work embodying the main epistemological ruptures that have played important roles in adding to the general understanding of the mechanisms of speech. Although we gladly acknowledge that this latter goal has been pursued through vast amounts of experimental and theoretical work worth writing about, its absence here is justified by the relatively narrow focus of our interest as well as by our wish to keep the chapter concise.

1.2.1 Speech production theories

Over the years, we have witnessed important theoretical achievements in the field of speech production (for a review, see Carré and Mrayati, 1990). Chiba and Kajiyama (1941) proposed the *wave propagation theory* of speech production in 1941. They showed that the speech sound is produced by one-dimensional acoustic wave propagation in the vocal and nasal tracts. Standing waves of flow velocity and pressure exist in the vocal tract across several resonant modes. Figure 1.2 illustrates a basic case of the application of this theory, namely the natural (i.e., cylindrical) tube case. This theory is well established and the standing waves it predicts have been confirmed by measurements (e.g., Firth, 1986).

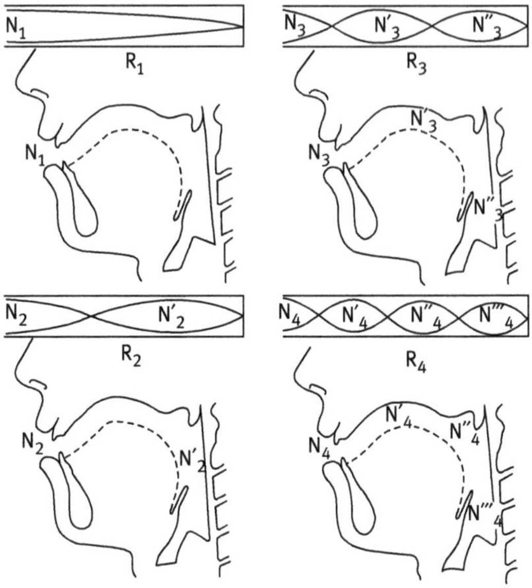

Fig. 1.2: Standing waves in the vocal tract (after Chiba and Kajiyama, 1941, Figure 93, p. 147, with permission by The Phonetic Society of Japan).

Source filter theory was a great step forward in the formulation of the speech production process and its models. In his book, Fant (1960) elaborated on the basic principles of this process providing explanation for a number of phenomena, principles that have been the foundation of models developed by other investigators. Figure 1.3 provides a simplified illustration of the source filter process for voiced speech sounds. In the time domain, the source output is a periodic train of glottal pulses generated when the airflow from the lungs is periodically interrupted by the glottis with an inter-pulse interval $T0$. In the frequency domain, the source has a spectrum showing the harmonics of the voice fundamental frequency $F0 = 1/T0$. The vocal source impedance is considered significantly larger than the tube impedance (Flanagan, 1972; Mrayati et al., 1976) and, therefore, the effect of the behavior of the source on that of the tube will be negligible with the result that in such a scheme there will be no coupling between the vocal source and the vocal tract. Assuming that both the vocal source and the vocal tract are linear systems, the spectrum of the radiated signal will be the product of the harmonic line spectrum $S(f)$ of the source and the transmission spectrum $T(f)$ consisting of the tube's main (1st, 2nd, 3rd…) resonance modes, that is the formants. The source filter theory is well adapted for studying static configurations, both one-by-one or a succession thereof.

The *perturbation theory* has been successfully used to study the relation between a small variation of an area function and the corresponding variation of acoustic parameters (Ungeheuer, 1962; Schroeder, 1967; Mermelstein, 1973). Fur-

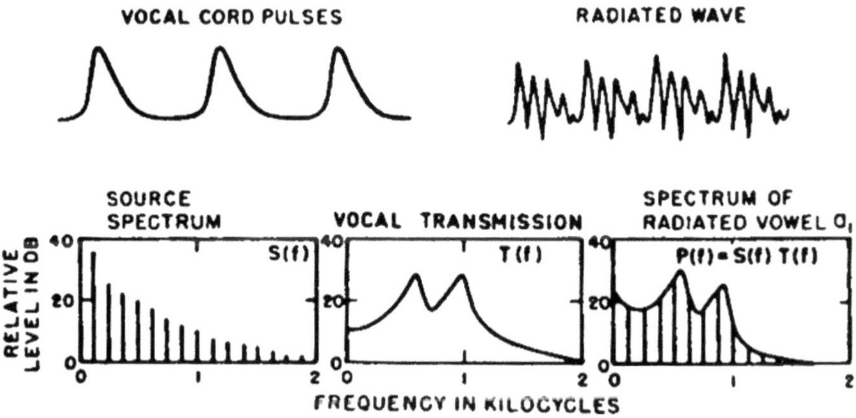

Fig. 1.3: The source-filter concept (from Fant, 1960, Figure 1.4-8, p.74, with permission by de Gruyter, Inc.).

ther investigations of the perturbation theory have led to the concept of *sensitivity function* (Fant, 1967, 1975a, 1980; Fant and Pauli, 1974; Mrayati and Carré, 1976; Wood, 1979). For a given formant and a position x along the length of the tube, the sensitivity function expresses the relation between a small local change in the cross-sectional area of the area function $A(x)$ and a change in the formant frequency or bandwidth. The sensitivity functions thus identify parts of the vocal tract that are "sensitive" for changing formant frequencies.

Figure 1.4 displays sensitivity functions for different Arabic vowel configurations (Wood, 1979). Sensitivity functions are generally calculated for positive local perturbation of $A(x)$, that is for local increases in the area function. As an example, we note that F_1 of the vowel [u] rises as the area at the lips increases.

The sensitivity function for perturbations over the length of the tube, rather than over its local cross-sections, can also be calculated (Fant, 1975a). These functions are a measure of the relative dependence of the particular resonance mode on various parts of the area function. This is the best definition we have of formant-to-cavity affiliations. Sensitivity functions relate formant frequency changes to *small local perturbations*. Inversely, area perturbation functions can be calculated in order to obtain a specific formant frequency change (or no change at all). These functions will be presented formally in Chapter 2.

Figure 1.5 illustrates the variation of the first formant obtained by applying an anti-symmetrical perturbation to a uniform area function, in proportion to the sensitivity function curve corresponding to the first formant. It shows that only the first formant changes its frequency, with all other formants remaining stable (Schroeder, 1967). Conversely, applying a symmetrical perturbation about the

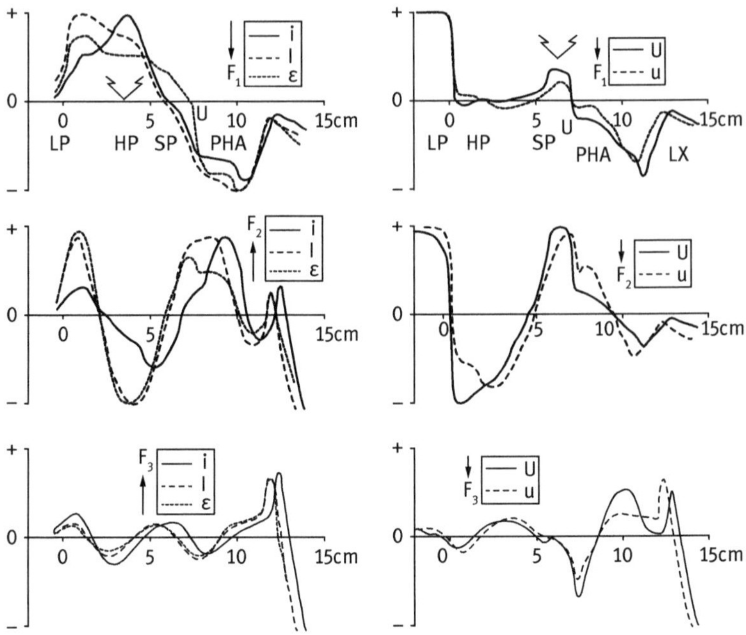

Fig. 1.4: Example of F_1, F_2, F_3 sensitivity functions for different vowel configurations (after Wood, 1979, Figure 3, p. 28, with permission by Elsevier Publishing Co.).

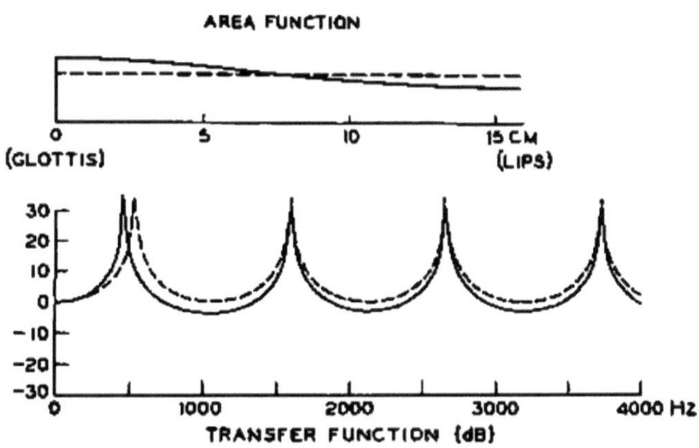

Fig. 1.5: Example of the application of perturbation theory in the three formant clamping case (F_2, F_3, F_4) when F_1 only is changed (from Schroeder, 1967, Figure 3a, p. 1005, with permission by the Acoustical Society of America).

midpoint of the uniform tube has almost no influence on the formant pattern. Heinz (1967) generalized the work of Schroeder on a uniform tube to arbitrary area functions.

Perturbation theory, as presented above, is well adapted to the study of quasi-static configurations. But changes in the shape of the vocal tract during speech production are larger than the small perturbations investigated in the above studies, thereby making it necessary to explore different computational means, in order to make these theories applicable to real-world speech.

Finally, the *quantal theory* of speech, proposed by Stevens (1972), was another important achievement in the theoretical formulation of the speech production process. Stevens postulated that the articulatory-acoustic relations are quantal in the sense that, even if the articulatory parameters vary continuously, the acoustic pattern could change from one (quasi-)steady-state to another. Figure 1.6 shows such a relation. In region II, large acoustic change happens for small shifts in articulation. In regions I and III, the acoustic parameter remains relatively stable when large variations are made in the articulation. It is suggested that "the existence of relations of this type between articulation and sound output has certain implications with regard to the inventory of articulatory and acoustic properties that are used as codes for signaling linguistic information" (Stevens, 1972). Two examples confirming the quantal theory are (1) the sudden acoustic change when a constriction gradually increases, and (2) the presence of an acoustic stability during the process of producing certain vowels.

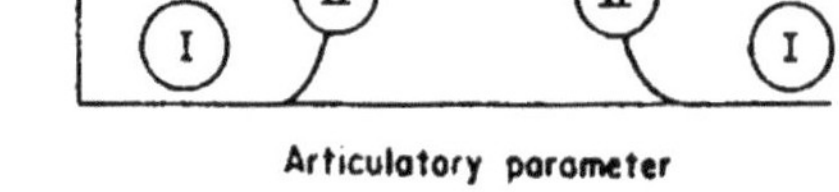

Fig. 1.6: Schematized relation between an acoustic parameter of speech and a parameter that describes some aspect of the articulation (from Stevens, 1972, Figure 1.6, p. 52, with permission by McGraw-Hill Co.).

1.2.2 Speech production data

1.2.2.1 Articulatory data

Fant's theory was tested using X-ray data on vowels generated by a Russian speaker (Fant, 1975a). These pictures make it possible to see the shape of the vocal tract in its entirety, from the larynx cavity to the lips. Analysis of the data by Sundberg et al. (1987) transforms the sagittal view to cross-sectional areas. To investigate first static, but even early on essentially dynamic speech production, numerous studies have used X-ray (Wood, 1979), palatography (Butcher and Weiher, 1976; Recasens, 1984; Farnetani et al., 1985), and more recently MRI (Winkler et al., 2006; Narayanan et al., 2014) methods. Wood (1979) reports on vocal tract area functions estimated from 38 sets of X-ray vowel articulation in English and Arabic speech, recorded by him as well as others. He shows that three main constriction locations exist: those situated along the hard palate (/i/, /e/, /ɛ/, the soft palate (/u/), or in the lower pharynx (/ɑ/, /a/, /ae/). Another constriction place is found in the upper pharynx (/ɔ/, /o/), a location approximately in the midpoint between the soft palate and the upper pharynx. He concludes that each location is appropriate for defining the qualities of a given class of vowels, thus showing that these vocalic classes confirm the quantal nature of vowel articulation. His four quantal regions are shown in Figure 1.7. Symmetrical and asymmetrical shapes of the area functions can be observed on each panel of the figure. Furthermore, the tongue musculature is well adapted for such deformations (Wood, 1979, his Figure 1).

Using synchronized electropalatographic and audio recordings of Catalan and Italian VCV sequences, Recasens and Farnetani (Recasens, 1984; Farnetani et al., 1985) identified vocalic contexts with low and with high degrees of coarticulation – effects that previously could not have been observed.

Researchers in Naraynanan's laboratory reported on a database of synchronized audio, midsagittal MRI, and electromagnetic articulograph (EMA) recording of ten English-speaker subjects (Ramanarayanan et al., 2013) and observed consistently identified places of articulation and regularly spaced vocal tract regions, in agreement with our model described in Chapters 3 and 4.

1.2.2.2 Acoustic data

Early on, it was recognized that electronic circuits and, later, digital computer simulations could lead to spectrographic representations of the speech signal (Koening et al., 1946; Oppenheim, 1970). Using this representation, vowels and consonants can be described in terms of resonant frequencies of the vocal tract, i.e., formants, and noting this fact has led to a large number of studies. Among these, we should mention the seminal work of Peterson and Barney on vowels (1952) that we mentioned earlier, showing that the representation of vowels in the F_1-F_2 plane is, to a large extent, speaker-dependent and displays large differences between male, female, and child recordings. Most importantly, this study demon-

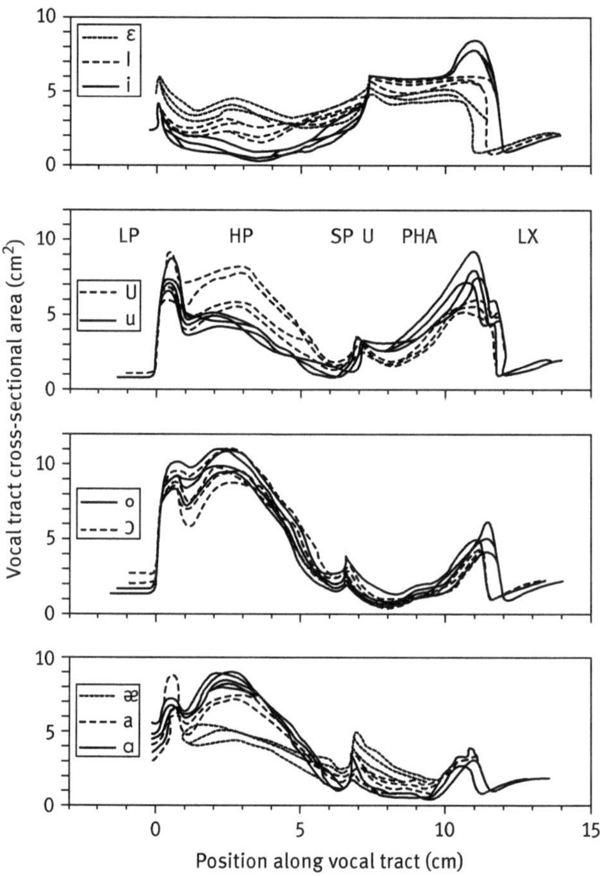

Fig. 1.7: Area functions for English vowels (after Wood, 1979, Figure 1, p. 26, with permission by Elsevier Publishing Co.).

strated that all vowels are located inside the acoustic plane we now call vowel triangle (see Figure 1.8).

1.2.2.3 Phonetic and phonological data

Sound inventories of world languages represent valuable data containing information on general trends and allowing tests of various hypotheses on phoneme structuration and representation. Diachronic evolution (i.e., sound changes that show the time course of languages traversing an evolutionary process) may also provide explanation for the structuration of the phonetic components of a given language.

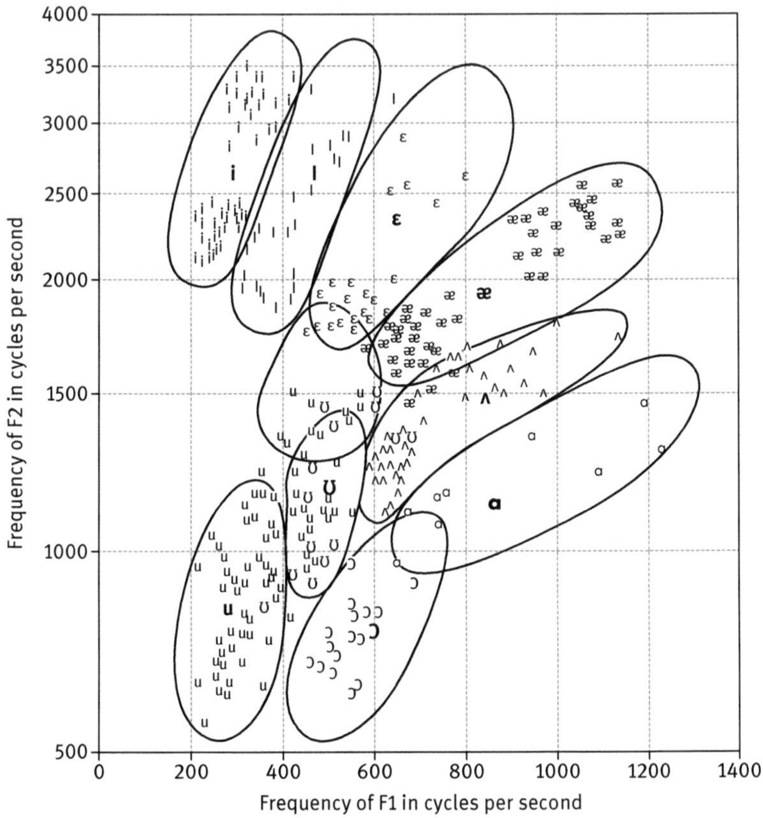

Fig. 1.8: Vowels spoken by men, women, and children, represented in the F_1-F_2 plane (after Peterson and Barney, 1952, Figure 8, p. 182, with permission by the Acoustical Society of America).

Among inventories, we refer to two well-known corpora, the *Crothers corpus* (Crothers, 1978), and the *UCLA Phonological Segment Inventory Database UPSID* (the UCLA Phonological Inventory Database, Maddieson, 1984). The first describes vowel systems and their main characteristics in 209 languages, while the second contains the description of vowels and consonants in 317 languages. The number of phonetic segments in UPSID varies between 11 to as many as 141, with those in the middle of the distribution containing between 20 and 37 segments. A tentative relationship between size and structure is also described. Other databases containing syllabic or semi-syllabic elements are also available. Among those, databases on spontaneous speech signals, such as the Sankoff–Cedergren French corpus (Sankoff et al., 1973; Bourdeau et al., 1990) are likely to be useful for visualizing the frequency of use of a particular phonetic component or combinations of phonetic components.

1.2.3 Speech production modeling

The data presented in the preceding section has generated a number of theories allowing simple and straightforward representations of vocal tract shapes. Specification of the area function in the form of an essentially continuous graph of cross-sectional areas that extend from the glottis to the lips yields a useful practical base for the calculation of the acoustic response, although the method is not ideally suited for a systematic description of the articulatory process. Several models of speech production have been proposed for providing insight into the relation between area-function and formant behavior or, in more general terms, between the articulatory process and its acoustic representation. One advantage of models is that they describe speech production by way of a small number of parameters. In the following section, we will review the main production models with their nomograms – functions that relate formant frequency changes to variations of model parameters one at a time and that permit viewing the dynamic behavior of the model.

1.2.3.1 Area function modeling

On the basis of data mentioned in the preceding section several models have been proposed. As their main characteristic, they consist of a certain number (2, 3, 4, etc.) of successive cylindrical cavities that approximate a given configuration of the vocal tract. The parameters of these models specify the length and the width of these cavities that represent, as a whole, the place and the degree of tongue constriction. In addition, these articulatory models are controlled by articulatory parameters, such as tongue height, tongue body, tongue tip, etc.

As early as in 1955, Stevens and House (1955) proposed a three-parameter model (place of articulation, degree of constriction, and lip parameter). Fant's model (1960) differs from this one only in its details (Figure 1.9). He provided nomograms relating the first five formant frequencies to the dimensions of the cavities and to the location and area of the constriction. In such nomograms, dynamic deformation of the area function is obtained by longitudinal displacement of the constriction. However, this does not mean that the specific longitudinal displacements in question are actually those observed in speech.

Using X-ray images taken during the production of ten vowels by five English speakers, Harshman et al. (1977) performed a factor analysis of the recorded

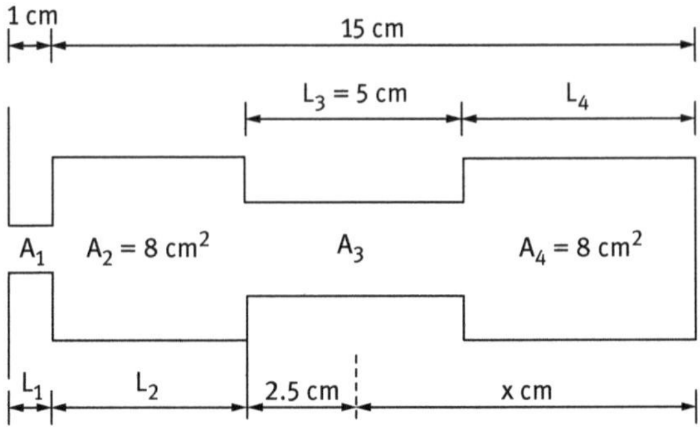

Fig. 1.9: The model after Fant (1960, Figure 1.1-2, p. 19, with permission by de Gruyter, Inc.).

tongue movements. The results show that tongue shapes can be satisfactorily described by two underlying main factors of tongue gestures around the neutral position: front rising and tongue rising. Analysis of X-ray data of French-speaking subjects reached a similar conclusion although, instead of two, it identified four factors, thereby allowing a finer definition of the gestures (Maeda, 1990, see Figure 1.10). These two studies have uncovered *synergetic* movements, symmetrical as well as anti-symmetrical, applied to specific regions along the tongue. From the point of view of dynamics, these data show that the vocal tract changes its shape by way of a spread of transversal deformations rather than as a longitudinal displacement of the place of the constriction.

1.2.3.2 Articulatory modeling

The articulatory model by Coker et al. (1973) describes the vocal tract in terms of seven parameters: the position of the tongue body (X, Y), the lip protrusion (L), the lip rounding (W), the place and degree of tongue tip constriction (R, B), and the degree of velar coupling (N, see Figure 1.11).

The articulatory model by Lindblom and Sundberg (1971) is controlled by six parameters: labial height, labial width, jaw, tongue tip, tongue body, and larynx height. A special feature of this model is that it reinterprets the classical notion of tongue height. Mermelstein (1973) devised a different articulatory model, one which describes the vocal tract by means of variables specifying the position of the jaw, tongue body, tongue tip, lips, velum, and hyoid. Ladefoged and Lindau (1979) and Maeda (1979) proposed an articulatory model based on the statistical analysis of tongue shape (see Figure 1.11). To describe a speech event, the tongue is regarded either as a composite of two independent parameters (front rising and back rising) or as a composite of independently controllable systems, such as the jaw, the tongue-body, the tongue dorsum, and tongue tip.

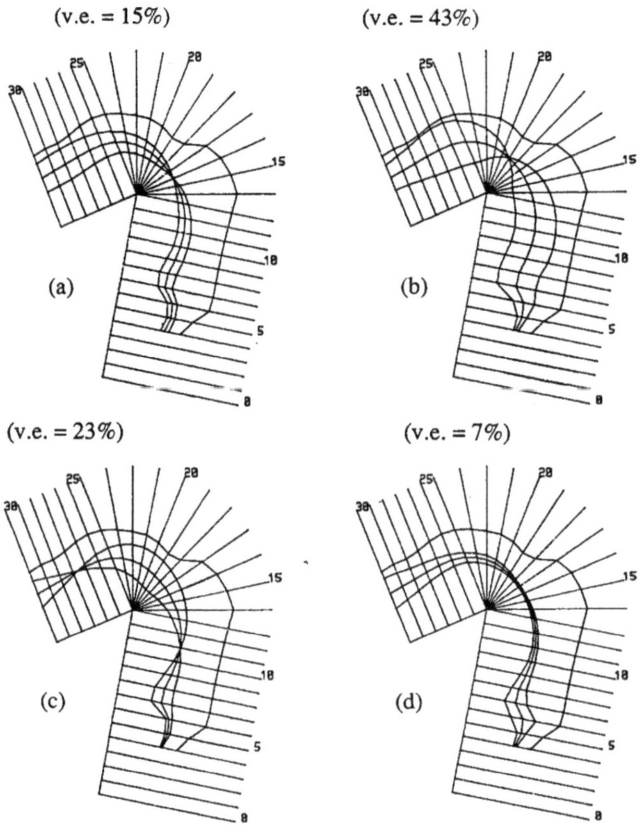

Fig. 1.10: Decomposition of the tongue contour into four linear components (jaw (a), tongue body (b), tongue dorsal (c), and tongue tip (d); (Maeda, 1990, Figure 3, p. 139, with permission by Springer Publishing Co.).

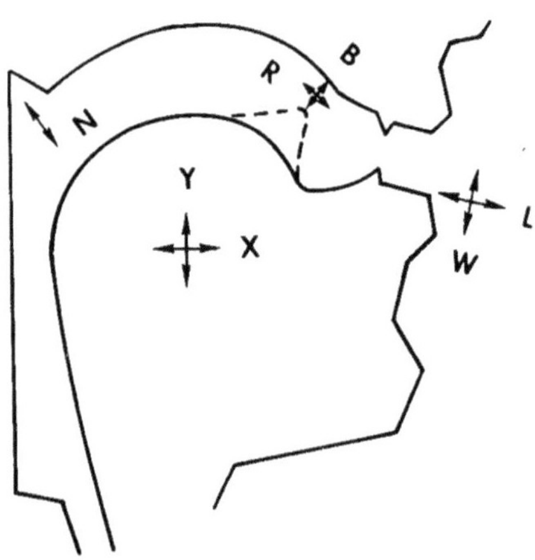

Fig. 1.11: The Coker Model (from Coker et al. 1973, Figure 2, p. 294, with permission by IEEE).

1.2.3.3 Spectral modeling

The first synthesizer to test the formant parameters visible on spectrographic data was the pattern playback apparatus developed at Haskins Laboratories (Cooper et al., 1951). To obtain synthesized speech signals for experiments with nonsense syllables, a light was shone through a slowly advancing transparent film on which formant patterns corresponding to the syllables in question had been hand-painted. This equipment was extensively used in perceptual tests that demonstrated, for example, the role of formant change in plosive consonant-vowel syllables, and that eventually led to proposing the notion of *locus* (Delattre et al., 1955)[2].

Around the same time period, it was recognized by researchers, e.g., Lawrence (1953), Fant (1956) and Flanagan (1957), that the transfer function of the vocal tract actually embodies the concept of a formant synthesizer and using it can result in a more natural speech synthesis. Such formant synthesizers, for example, the PAT (Parameteric Artificial Talking Device) developed in Edinburgh by Lawrence (1953), and the OVE II synthesizer developed in Stockholm by Fant and his colleagues (1962), were singularly well adapted for speech research – especially that on speech perception. Furthermore, predictive coding techniques based on an all-pole model of the transfer function were shown to be able to become a tool for automatical extraction of formant frequencies from the speech signal (see e.g., Atal and Hanauer, 1971).

1.3 Speech perception

1.3.1 A short history of the study of speech perception

Apart from considering the extensive work on speech over the centuries in Arabic civilization (Bakalla, 1982), approaching the study of speech from a historical perspective one is struck by its young age in the Western civilization. From the classic European period up until more than 150 years ago, speech was regarded just as a means for the expression of thoughts put into words – λόγοσ (logos). According to Aristotle, speaking was a natural way of giving manifestation to thoughts, reason, desires – as language became part of the arts of logic, rhetoric, and poetics (McKeon, 1946). Aristotelian thinking of language and verbal communication remained central to Western philosophy well into the early 1800s, a period that witnessed the birth of the science of linguistics after Humboldt (1999) postulated that

2 The locus of a transition is specified by the formant frequencies at the end of a VC or at the start of a CV transition.

language was a rule-governed system. But, not incidentally because of the high standing of German thinking and science throughout that era, the study of *speech* was not separated from that of *language* – "Sprache" is the German word for both. Even at the end of the 19th century, despite formidable advances in physiological and physical research, Wilhelm Wundt (1874, p. 610 ff), considered speech as one among the modes of expression of affect and ideas, at par with gestural communication – a mode he considered almost as effective. Expression of affect and ideas is the outcome of the human drive of reaching out, but the object of expression is present even when not externalized: thoughts, wishes, dreams – all within the realm of language as words, concepts. Perception of speech fits into his principle of apperception: recognizing a sensory input as something reflecting a resemblance to an earlier, internalized experience (Wundt, 1874, p. 305 ff)[3] – in other words, perception of spoken language allows the acoustic form to immediately access the reference.

Maybe because in Romance languages, in French in particular, the words referring to speech and language are not identical ("parole" vs. "langue") – it was a French native speaker Swiss linguist, Ferdinand de Saussure, who apparently separated the two. In his treatise, posthumously published by two of his co-workers in 1916 (de Saussure, 1995), he distinguished between the linguistic concept and the spoken word, referring to the first as the *signified* content and the second as the *signifier* form. Because, independently from this distinction, he considered speech and language a unified rule-governed system, de Saussure is credited to be the originator of linguistic structuralism, a conceptual framework quickly absorbed by groups of linguists interested in the structural analysis of language, including that of speech sounds. Among these groups was the "Group of Prague" that included Trubetzkoy, one of the forefathers of phonology (1969), and Roman Jakobson who, after starting as a scholar of Russian linguistics, adapted the structuralist framework to the analysis of phonemes within and across languages. When he moved to the U.S., Jakobson took Claude Shannon's work on communication theory (Shannon, 1948) as the foundation for his groundbreaking work on the distinctive features of phonemes (Jakobson et al., 1951).

As a consequence of de Saussure's signifier-signified dualism, or at least on a parallel course with it, the study of the perception of speech has also come to recognize the two components of language. While one of the earliest reports on a controlled experiment on speech perception by Bagley (1900) was directed at the recovery of meaningful words (recognition of incomplete words recorded on Edi-

3 At the time of the present book's publication, Wundt's *Grundzüge der physiologischen Psychologie*, 2nd ed., was available at https://archive.org/details/grundzgederphys15wundgoog/.

son's cylinder) – and thereby at testing Wundt's apperception principle – investigations directed at the perception of Jakobson's distinctive features in nonsense syllables at Haskins Laboratories addressed the question how the listener deals with the acoustic form of speech (see, e.g., a study by Liberman and colleagues (1954). At Bell Laboratories, Flanagan and his colleagues recognized early on the importance of the ear's sensitivity for the resolution of various acoustic parameters of speech, such as static and dynamically changing formant frequencies, formant bandwidths, and levels (Flanagan, 1955b, 1955a, 1956). However, he was also aware that these psychophysical evaluations of the speech signal relied on the listener's judgments that were comparative, that is *relative*, and thus were different from the *absolute* judgments man needs to make when the task is understanding speech (Flanagan, 1972). This latter task necessitates much more than the sensory processes needed for decoding the acoustics of the speech signal: understanding words, phrases, sentences, and longer discussions taps into the person's experience with the language of the speech heard. This experience allowed him to store a vocabulary and to learn the syntax, and also to retrieve the relevant items from the memory storage efficiently enough to recognize words even under very unfavorable acoustics. Perception of speech is therefore a sensory-cognitive process. Its broader, knowledge-driven cognitive-linguistic aspect – the *deep level* – involves a host of cognitive activities, such as learning, long-term and working memory, adaptation, information processing, hypothesis testing, and decision making. The stimulus-driven *surface level* sensory aspect of speech perception is often regarded as a psychophysical task (see the book edited by Schouten, 1987), whether involving simple (Repp, 1987) or complex (Macmillan, 1987) decisions.

This sensory aspect can explain a number of phenomena encountered in running speech, the most important of which is the way our auditory system deals with the multi-dimensionality of the speech signal. A sub-aspect within it relates to the distribution of the parameters within each of the dimensions and it is used to predict discrimination of speech sounds of a given type. For example, in a set of vowels, discrimination is first based on the acoustic contrast between members of the set but it is also influenced by the frequency of occurrence of each member within the language of the stimulus. The sensory aspect of speech perception also brings into focus the overriding role of dynamics because, as all sensory systems being especially sensitive to changes, the auditory system responds particularly fast to changes occurring along any dimension of the speech signal.

As Repp (1987) pointed out, speech perception is inherently *relational*: while on the phonetic surface level perception depends on the listener's inner representation of the phonological structure, on the deep, linguistic level it is the outcome of interactions with mental structures formed through experience and/or genetics. In other words, when listening to speech, it is not the stimulus input that is perceived but its relation to stored phonetic and cognitive-linguistic knowledge. Perceived speech is simply the best fit between the input and the stored knowledge, and the accuracy of the perception is the goodness of this fit. This relational concept though could be pushed to an extreme position – by asserting that there are no objects of perception, only objectives to arrive at an agreement between the world and the organism (Kluender et al., 2003). According to such an empiricist view, perceiving the true state of the world is impossible and thus one cannot recover or reconstruct the distal state of speech, that is, the gestures that created it. This view is the diametrical opposite of the *direct* ("naive") *realism* advocated by Gibson (1972), although actually proposed in the Third Book of Aristotle (McKeon, 1946). Its epistemological opposite, the *indirect* ("representational") *realism*, asserts that what we perceive is not the external objects and events but our interpretations of a sensory input derived from the real external world. Among its first adepts, we recognize John Locke (see Uzgalis, 2016) who proposed perception of qualities – primary that perceive phenomena without needing further explanation, and secondary that refer to traits of objects, rather than the objects themselves.

There have been two opposing theoretical arguments on what constitutes, for de Saussure's "signifier" or Repp's surface component of speech, the actual structure on which perception of the phonetic form is based. A number of speech scientists have espoused the *indirect*, auditory-representational view of speech perception (e.g., Diehl et al., 2004). The principal exponent of *direct* realism for speech has been Carol Fowler (2016), although her line of thinking has been shaped by the *motor theory* of speech, first proposed in the 1950s by Alvin Liberman, Franklin Cooper, and their colleagues at Haskins Laboratories (1967) and later revised by Liberman and Mattingly (1985). The Haskins group asserted that what we perceive in speech goes beyond decoding its acoustic form – actually it is the articulatory gesture that generated the acoustic signal. The motor theory of speech perception has two major claims. The first is that the objects of speech perception are the intended phonetic gestures of the speaker, represented in the brain as invariant motor commands, and these gestural commands are the physical reality underlying the traditional phonetic notions that provide the basis for phonetic categories. The second claim of the motor theory is that speech perception and speech production must be intimately and genetically linked, such that there is actually a specialized mode of perception reserved for speech. In this perceptual mode a special code constitutes the link between the articulatory gesture and the phonetic structure.

One of the main differences between the two theories has been the way they explain the feature of *invariance* – the fact that we recognize a given property of the speech signal across acoustically different forms it can take, such as the consonant [t] across the CVC syllable list of /ati/, /uto/, /ete/, etc. The auditory representational argument states that it is the dynamic spectral pattern, consisting of the formant transitions to and from the consonant and of the plosive burst, that will uniquely define [t], and it is this pattern on which the listener bases his perceptual decision. In contrast, the motor argument posits that the different signal properties of the list share the underlying alveolar closure gesture and it is this gesture that the listener extracts from the signal as his or her decision variable. According to Fowler, the speech signal perceived by the ear immediately evokes the articulatory movements of the talker and it is directly perceived as the event that generated the signal, not necessitating the translation of the indirect auditory form leading to the recovery of the event. The direct perception theory offers the advantage that both the invariance and the dynamics of the speech signal are inherent properties therein. In addition, neurophysiological evidence demonstrated the existence of activity in motor and premotor cortical areas when listening to speech, activity localized to brain areas specialized for movements in the very regions of the vocal tract where the perceived speech gesture originated (Wilson et al., 2004; Pulvermuller et al., 2006; D'Ausilio et al., 2009).

1.3.2 Models of speech perception

The proof of any theory is whether, when using the same stages in the same order described therein, the results it implies, or outright predicts, can be reproduced by a computational model. For theories of speech perception the bar is high: a model would have to essentially repeat by live voice, or transcribe, what a talker just said. Of course, in our time this is routinely done by automatic speech recognition (ASR) machines, the most successful of which can achieve nearly perfect performance. But do these machines actually go through the stages of a theory that depict what we presently know about the way the human auditory and central nervous systems achieve speech perception? Unfortunately, even if some of a given machine's stages may have been inspired by up-to-date knowledge of the human systems, the answer is negative, although not because of the machine's fault. It just happens that, while the ASR models' acoustic analysis stages usually replicate the peripheral auditory system to an acceptable degree and subsequent stages can recover some (even if not all) phonetic patterns, we do not know sufficiently well the mechanisms inside the central nervous system that stand behind recognition of words, and even less those responsible for the recovery of meaning. Thus, stages of ASR models following acoustic analysis often consist of elaborate statistical processes (hidden Markov models, generalized Gaussian models, neural network models, etc.), most of which necessitate preliminary learning of the talker's individual speech characteristics. Also, only very few of the models we have reviewed use, in their analysis of the acoustic signal, estimates of vocal tract resonances. Below, we present two classic models selected for their human-inspired choices of analysis stages and the theoretical foundations behind them.

The first model is the one by Dennis Klatt (1979), a scientist-engineer who managed to put forward both a system of speech synthesis, very popular for over a decade-or-so after its creation, and a system of speech perception. His perception machine is based on two back-to-back systems, the (trans)SCRIBER and the LAFS, Lexical-Access-For-Speech. On the basis of psychoacoustic knowledge, SCRIBER derives a spectral representation of the signal and passes the series of spectra through a network of sequences of expected static spectra, each of which representing a transition between phonetic segments. It bypasses phonetic segmentation in favor of diphones, and performs both time and talker normalization. LAFS works in parallel with SCRIBER, that is, it generates lexical hypotheses directly from spectral representation of the signal without first recognizing phonetic segments. It is based on acoustic-phonetic knowledge and on a word boundary phonology, precompiled into a decoding network of expected series of spectra and possible sequences of words in the sequence lexicon; it therefore avoids early phonetic decision errors that would become word errors when searching in the lexicon. Thus, LAFS has lexical representations for optimal search, performs phonological recoding of words in sentences, deals with initial phonetic representation errors, and interprets prosodic cues to lexical items and to sentence structure. Klatt made the software for his system available to users but, because of his untimely passing, he did not subject it to the rigorous performance testing he had intended.

The second system for speech perception we want to present is James McClelland's and Jeffrey Elman's TRACE model. Two versions of the model, with identical philosophy – connectionism – but different objectives, have been developed. TRACE I (described in Elman and McClelland, 1986) was designed to process real speech, so it had to possess knowledge of exactly what features must be extracted from the speech signal, of what the differences in duration of different features of different phonemes were, and of how to cope with features and feature durations that vary as a function of context. Although TRACE II (McClelland and Elman, 1986) stopped at word perception, it went far into the cognitive realm, primarily to account for lexical influences on phoneme perception and for direct recognition of words, while also revealing some aspects of phoneme perception as being natural consequences of the TRACE framework[4]. The model is based on principles of interactive activation: Information processing takes place through the excitatory and inhibitory interactions of a large number of simple processing units, each of which works continuously to update its own activation on the basis of the activations of other units to which it is connected. It is called TRACE because this network of units defines a dynamic processing structure called "the Trace," serving as not only the perceptual processing mechanism but also the working memory of the system. The model is built from three interconnected *levels*, and at each level there are several *units* corresponding to individual properties or items. Each unit has a resting activation level; each connection has a numerical weight that determines how fast it conducts information and thus how much information is allowed to transfer over a given time interval. Processing takes place through the excitatory and inhibitory interactions of the units. Units on different levels that are mutually consistent have mutually excitatory connections, while units on the same level that are inconsistent have mutually inhibitory connections. All connections are bidirectional.

The *feature level* consists of several banks of feature detectors, one for each of the various speech sound dimensions. Each bank is replicated over several successive moments in time – called time slices. The *phoneme level* consists of *units* and detectors, one for each phoneme. One copy of each phoneme detector is centered over three time slices but each phoneme unit spans six time slices, so units with adjacent centers span overlapping ranges of detector slices. Phoneme units are also context sensitive (i.e., they can decode coarticulated/co-produced phonemes). At the *word level*, there is one detector for each word, one copy of which is centered over every three feature slices, but each word detector spans a stretch of as many feature slices as the entire length of the word requires. Speech input enters the model at the feature level. It is presented in successive slices to the feature units, in the form of a pattern of activation unfolding in time. All the units at all three levels are continually active during the processing. At each time slice a new input slice is accepted by the first time slice of the unit, while the previous input frame is shifted to the right. The entire network of units is called "the Trace," because the pattern of activation left by a spoken input is a trace of the analysis of the input at each of the three processing levels. It is dynamic, since it consists of activations of processing elements, and these processing elements continue to interact as time goes on.

4 A Java-version of TRACE has been implemented at the University of Connecticut and, at the time of the present volume's publication, was available online at http://magnuson.psy.uconn.edu/jtrace/.

The model's major achievement is that distinction between perception and (primary) memory is completely blurred: the percept unfolds inside the same structures as those that perform the task of working memory – memory that allows perceptual processing of older portions of the input to go on while newer portions enter the system. In addition, the continuing interaction between the units within and across levels permit the model to incorporate the appropriate context effects and also to directly account for certain aspects of short-term memory, such as the fact that more information can be retained for short periods of time if it hangs together to form a coherent whole.

Despite the incontestable accomplishments that both the Klatt and the TRACE models represent, they do not address perception of speech connected to the articulatory origin of the input. That approach, speech perceived as gestures, was adopted by the motor theory of speech perception mentioned earlier but, unfortunately, has not been converted into a model. Although our theory and model described in Chapters 3 and 4 do address the issue from the point of view of speech production, and Chapters 5, 6, and 7 show results that confirm the validity of that theory and model in the perceptual domain, building a model of speech perception based on gesture representation will necessitate another undertaking at some point in the future.

1.4 Conclusions

As stated in the Preface and the Introduction, our aim is to build a speech production *model*, based on acoustic tube properties. This chapter presented scientific work and models, not as an exhaustive survey but as a list of the work that inspired and helped us in the undertaking of our endeavor. Rather than as a body of results, theories, and models to pick elements from and to later assemble to our liking, we want to look back on them to see how they can become an aid for specifying the criteria that would need to be met to characterize a "good" *model* – the optimal model of production we want to build verified by its perceptual counterpart.

Accordingly, we think that a model of speech production based on acoustic theory would need to possess the following characteristics:

1. the capability of the model to describe natural phenomena as closely as possible (vowel and consonant production, compensation, coarticulation, vowel reduction, etc.);
2. the parameters of the model should be orthogonal and capable of producing all possible states of the modeled system;

3. control strategies consisting of simple but significant commands directed at a small number of model parameters;
4. nomograms expressing simple and easily interpretable relations between articulatory, acoustic, and phonetic domains, in order to provide the model with an instructive and explanatory tool of the speech production process;
5. the capability of exhibiting non-linear phenomena inherent in natural speech production;
6. the capability of predicting other phenomena such as ventriloquist speech, bite-block speech, formant clamping, speech following glossectomy, and male–female–child vowel invariance;
7. the capability of producing realistic-sounding synthetized speech signals to allow them to be verified in perception tests.

In a similar vein, a good model of speech perception would need to meet these criteria:

1. it should be consistent with knowledge of the sensitivity of the auditory system regarding static and dynamic parameters of the production model;
2. to efficiently deal with the dynamics of a speech signal, it should analyze the incoming speech signal simultaneously on several time scales;
3. it should have the capability of learning and recognizing static and dynamic formant patterns of phonetic units;
4. it should exhibit differential sensitivity for immediately signaling departures from expected phonetic patterns, phonetic or idiolectic dynamic changes in formant or fundamental frequency parameters;
5. it should ensure that the speech signal is intelligible even under unfavorable acoustic conditions, that is in noise and in reverberation;
6. it should have the capability of learning and recognizing individual speech characteristics of a given talker;
7. it should ensure intelligibility of speech by a target talker in the presence of speech by various numbers of non-target talkers, located either in a different or in the same spatial channel.

In the following chapters we will describe a novel approach viewing speech as a dynamic process, an approach not derived from data but solely from efficient deformations of the acoustic tube and the corresponding acoustic trajectories.

2 Perturbation and sensitivity functions

2.1 Introduction

In a deductive approach for investigating a complex system, *sensitivity analysis* offers a powerful tool. In this volume, this type of analysis is being used to understand the process of sound production by an acoustic tube, and to determine the relationships between certain input and output variables. It is also used to model this process by including efficient command parameters. In this chapter, we present the concept of the sensitivity function and recall the way it is derived for the acoustic tube. We also examine examples of this function's use for different tube structures and draw several informative conclusions that shed light on the process itself.

2.2 Sensitivity functions for process modeling

Sensitivity functions *SF* determine how *sensitive* a given output parameter of a *process* (or system) is to a change in a *control* parameter. In this chapter, the process is sound production by a cylindrical acoustic tube 18 cm in length. The input is an excitation by a source located at one end of the tube that is assumed to be always closed, whereas the output is the sound produced at the other end that is either open or quasi-closed. The output parameters are the resonant frequencies of the tube – the formants that derive from the spectrum of the sound produced; in this book we mostly work with the first two or three formants, that is with F_1 and F_2 and often also F_3. The control parameter is the deformation of the cross-sectional area of the tube $A(x)$ at the length coordinate x measured from the input end. The sensitivity function for this system is given by Fant and Pauli (1974) as

$$SF_i = [\Delta F_i/F_i]/[\Delta A(x)/A(x)] \tag{2.1}$$

where i is the formant number and $\Delta A(x)$ is an area perturbation leading to the variation ΔF_i of formant i.

The concept of sensitivity function has been routinely used in process- or system-modeling and in control theory (Saltelli, 2002). In modeling, it offers a methodology to understand and deduce both the features and the inherent behavior of the process, and to subsequently find how the model can achieve optimality. In control theory, it is used to determine efficient parameters for the control of the process. Sensitivity analysis can be useful for a range of purposes, including:

DOI 10.1515/9781501502019-003

- better understanding of the relationships between input and output variables in a system or model;
- model simplification – setting model inputs that have no effect on the output to a constant value, or identifying and removing redundant parts of the model structure;
- testing the robustness of the results of a model or system;
- searching for errors in the model (e.g., for unexpected relationships between inputs and outputs);
- finding regions in the space of input factors for which the model output is either maximum, minimum, or meets some criterion of optimality.

Sounds are produced using an acoustic tube that constitutes an element of a communication system, that is, a transmitter producing an acoustic code, where a receiver would co-adapt itself with the transmitter, in order to perceive these sounds. The sensitivity function can be used in modeling this production tube and in controlling the model for the production of specific sounds by way of efficient changes of the cross-sectional area, that is, by using efficient coding (Figure 2.1).

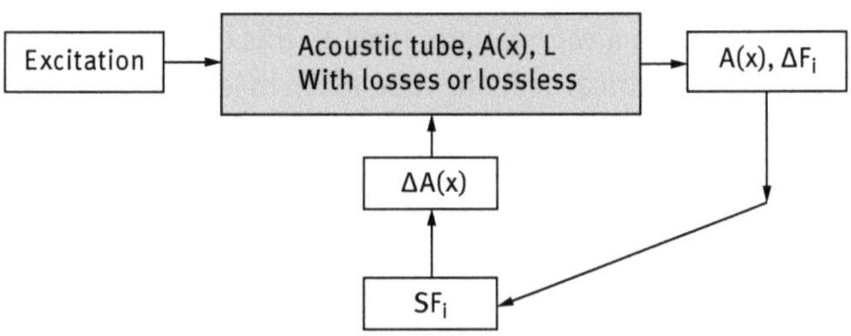

Fig. 2.1: Sensitivity function in modeling and control of a process.

2.3 Resonances and energy in an acoustic tube

An acoustic tube is characterized by its length L, by its shape defined by the area function $A(x)$ indicating the cross-sectional area at point $x \subset L$ regardless of whether its two ends are open or closed, and by the various energy losses that take place in the tube. Such losses are caused by distributed heat dissipation along the tube walls, viscosity, by the finite resistance of the excitation source, by the resistance of the tube wall's impedance (losses due to yielding walls or wall vibration), and by the resistive part of the radiation impedance. The contribution of these dissipative elements to the bandwidths of resonances F_1, F_2, F_3 has been analyzed in detail by Fant (1960, 1972), Fant and Pauli (1974), and Mrayati and Carré (1976). Losses through the wall impedance dominate the bandwidth B_1 when the frequency of the first formant F_1 is low. The surface losses determine B_2 and B_3 when the lip opening is very narrow. When the aperture is open, the radiation resistance is the main determinant of B_2 and B_3. Apart from the reactive component of the radiation impedance, which is to be reflected in a prolongation of the length of the tube to obtain its effective length, and the wall vibration effect, which is important in raising the low F_1 value, it has been concluded that the effect of losses on sensitivity functions *is not important*, as far as formant frequencies are concerned. In this book all unimportant sources of loss will be ignored.

When exciting the tube at its input end, an acoustic wave will propagate and will associate with standing waves at the resonant frequencies. For each formant we have spatial distributions of wave velocity, pressure, kinetic energy, and potential energy.

The kinetic energy KE_i and potential energy PE_i for each formant i are functions of pressure $P_i(x)$ and volume velocity $U_i(x)$ defined for each point x along the length of the tube as:

$$KE_i(x) = \frac{1}{2}\frac{\rho}{A(x)}|U_i(x)|^2 \tag{2.2}$$

and

$$PE_i(x) = \frac{1}{2}\frac{A(x)}{\rho C^2}|P_i(x)|^2 \tag{2.3}$$

where $A(x)$ is the cross-sectional area at point x measured from the input end of the tube, ρ is the density of air, and c is the speed of sound.

Figures 2.2, 2.3 and 2.4 show these reactive energies for the three formants. It is interesting to note that for the uniform tube the two energies have a π phase shift, so their sum will indicate a constant total energy all along the tube and their difference will result in an anti-symmetrical sinusoidal curve.

2.4 From perturbation theory to sensitivity function

It is essential to know the acoustic consequences of a deformation inflicted to any point of the tube and vice versa; *perturbation theory* offers the basic principle that defines this relationship. This theory has been successfully used to study the relation between a *small area function* perturbation and the corresponding changes of resonance characteristics (Ungeheuer, 1962; Mermelstein, 1973; Schroeder, 1967).

This concept was further developed, leading to the concept of *sensitivity function* (Fant, 1967, 1975a, 1980; Fant and Pauli, 1974; Mrayati and Carré, 1976; Wood, 1979). For a given formant, the sensitivity function for *local* perturbations of the area function $A(x)$ at any point of the tube relates such small local spatial variations to changes of formant frequency and bandwidth. These sensitivity functions provide information on how *sensitive* a formant frequency is to changing the shape of the tube at any point.

Schroeder (1967) was the first to use the general theorem of the perturbation theory by Ehrenfest (1916) to study the relation between perturbations in the area function of a *uniform acoustic tube* and the tube's resonant frequencies, that is:

$$\Delta(TE_i/F_i) = 0 \quad \text{or} \quad \Delta TE_i/TE_i = \Delta F_i/F_i \tag{2.4}$$

where TE_i is the total energy, F_i is the frequency, Δ (small change) represents an adiabatic perturbation, and i refers to one mode (resonance frequency).

We recall that at each of the resonance frequencies the total kinetic energy KE is equal to the total potential energy PE and consequently equal to half the total energy TE:

$$\int_{x=0}^{L} PE_i(x)dx = \int_{x=0}^{L} KE_i(x)dx = \frac{1}{2}TE_i \tag{2.5}$$

This procedure represents the derivation of a frequency change needed to restore equilibrium between potential and kinetic energies when an area is perturbed.

For a *closed-open* uniform tube, Schroeder arrived at several conclusions concerning *perturbations* of its area just near its uniform shape:
- a small *perturbation* of the area of this tube may change one formant while leaving the others unchanged;
- the area function is uniquely defined (to within an arbitrary constant) by complete specification of the poles and zeros of the *input admittance* measured at the open end whenever the tube length L is known;
- perturbation functions are anti-symmetrical about the midpoint of the tube;
- there exist an infinite number of perturbation functions applied *symmetrically* with respect to the mid-point of the uniform tube that will have negligible effect on the formant frequency pattern.

Heinz (1967) generalized the work of Schroeder involving a tube with uniform area functions to the general case of *arbitrary* area functions. Fant (Fant, 1967, 1975a; Fant and Pauli, 1974), using circuit theory, developed an elegant representation of the relation between a small variation ΔF_i of each formant F_i and a small variation $\Delta A(x)$ at *any section x* of the *area function* $A(x)$; specifically, they provided curves relating formant changes to *local perturbations* affecting a *specific area* rather than to *perturbation functions* over the *whole tube*. In other words, these functions can be used to evaluate the change in formant frequency caused by a deformation of the area of a given section of the tube, rather than its entirety. They were the first to call the curves representing this relation *sensitivity functions* and showed that the sensitivity value $SF_i(x)$ at each point is proportional to the difference between the kinetic energy KE and the potential energy PE divided by the total energy TE:

$$\frac{\Delta F_i}{F_i} = \frac{\int_{x=0}^{L} [KE_i(x) - PE_i(x)] \frac{\Delta A(x)}{A(X)} dx}{TE_i} = \int_0^L SF_i(x) \frac{\Delta A(x)}{A(X)} dx \tag{2.6}$$

$$SF_i(x) = [KE_i(x) - PE_i(x)]/TE_i \tag{2.7}$$

where i is the formant number, L the tube length, and $[KE_i(x) - PE_i(x)]$ is the Lagrangian.

Therefore, the sensitivity of a particular formant frequency $\Delta F_i/F_i$ to a change in tube cross-sectional area $\Delta A(x)/A(x)$ is equal to the difference between the kinetic energy KE and potential energy PE as a function of distance from the closed end of the tube divided by the total energy in the system. Two different methods leading to the same result have been developed. The first starts out from the criterion of the kinetic energy of a mode equaling the potential energy of the mode, as explained above. The second is concerned with the impedance condition for resonance, and it is obtained from Ehrenfest's theorem (Equation 2.8, as cited by Schroeder, 1967), which means that the frequency shift ΔF_i is proportional to the change in energy ΔE_i at one of the tube's modes:

$$\delta F_i/F_i = \delta E_i/E_i \tag{2.8}$$

Equation 2.9 further postulates that the *energy change* can be expressed as the product of the *radiation pressure* P_i *[sub-mode]*, which, except for a scalar factor, is actually equivalent to the *sensitivity function* for a specific mode, $(PE_i - KE_i) = -(KE_i - PE_i)$ and the area function deformation integrated over the total length L of the tube (Schroeder, 1967, p. 1004);

$$\delta E_i = -\int_0^L P_i(x)\delta A(x) dx \tag{2.9}$$

Applying the Cauchy–Schwartz inequality (2.10).

$$\int_0^L Pi(x)\delta A(x) dx \leq \left(\int_0^L Pi^2(x) dx\right)^{1/2} \left(\int_0^L \delta A^2(x) dx\right)^{1/2} \tag{2.10}$$

shows that when $\delta A(x) = P_i(x)$, the change in the energy is maximum:

$$\delta E_i = -\int_0^L Pi(x)\delta A(x)dx$$

Consequently, $\delta F_i/F_i$ is maximum and, therefore, F_i is *most sensitive to an area perturbation* $\delta A(x)$ *proportional to the sensitivity function* $P_i(x)$ for a specific mode. At this most sensitive condition Equation 2.8 becomes:

$$\delta F_i/F_i = \left[\int_0^L Pi(x)\delta A(x)dx\right]/E_i \tag{2.11}$$

And if $-P_i$ is replaced by $+(KE_i - PE_i)$, we get the same sensitivity equation as in Equations 2.6 and 2.7.

Thus, when the acoustic energy variations are proportional to the area function perturbations, maximum acoustic changes are produced and vice versa (Carré, 2004). A search for efficient acoustic changes means that we want to identify points along the tube where the area function will not be perturbed by deformations that would have no acoustic effect.

Figures 2.2–2.4 show, for each of the three formants, the spatial distribution of the parameters involved in the expression defining the sensitivity function for a uniform tube. These parameters are the *velocity* that determines the kinetic energy KE, and the *pressure* that determines the potential energy PE (see Equations 2.2 and 2.3 above). The total reactive energy TE is, naturally, the sum of KE and PE and it is *constant* all along the length of the uniform tube. As Equations 2.6 and 2.7 show, the sensitivity functions SF_i are given by the difference between KE and PE. All curves are sinusoidal.

It is interesting to note that for the uniform tube these sensitivity functions are anti-symmetrical and sinusoidal (see Figure 2.5). Consequently, the functions have an odd number of zero-crossings, namely 1, 3, and 5 for the first three formants, respectively. This property will be extensively exploited in Chapters 3 and 4.

In a similar manner, the contribution of each point along the tube to formant bandwidths has been calculated by Fant and by ourselves (Fant and Pauli, 1974; Fant, 1975b; Mrayati and Carré, 1976) but laying out the relation between $A(x)$ and bandwidths of the resonant modes of the tube would be beyond the scope of this book.

It is interesting to recall that there are two types of sensitivity functions relating area function perturbation to acoustic parameters. The first, the one presented above, deals with *transversal changes* of local areas and it is proportional

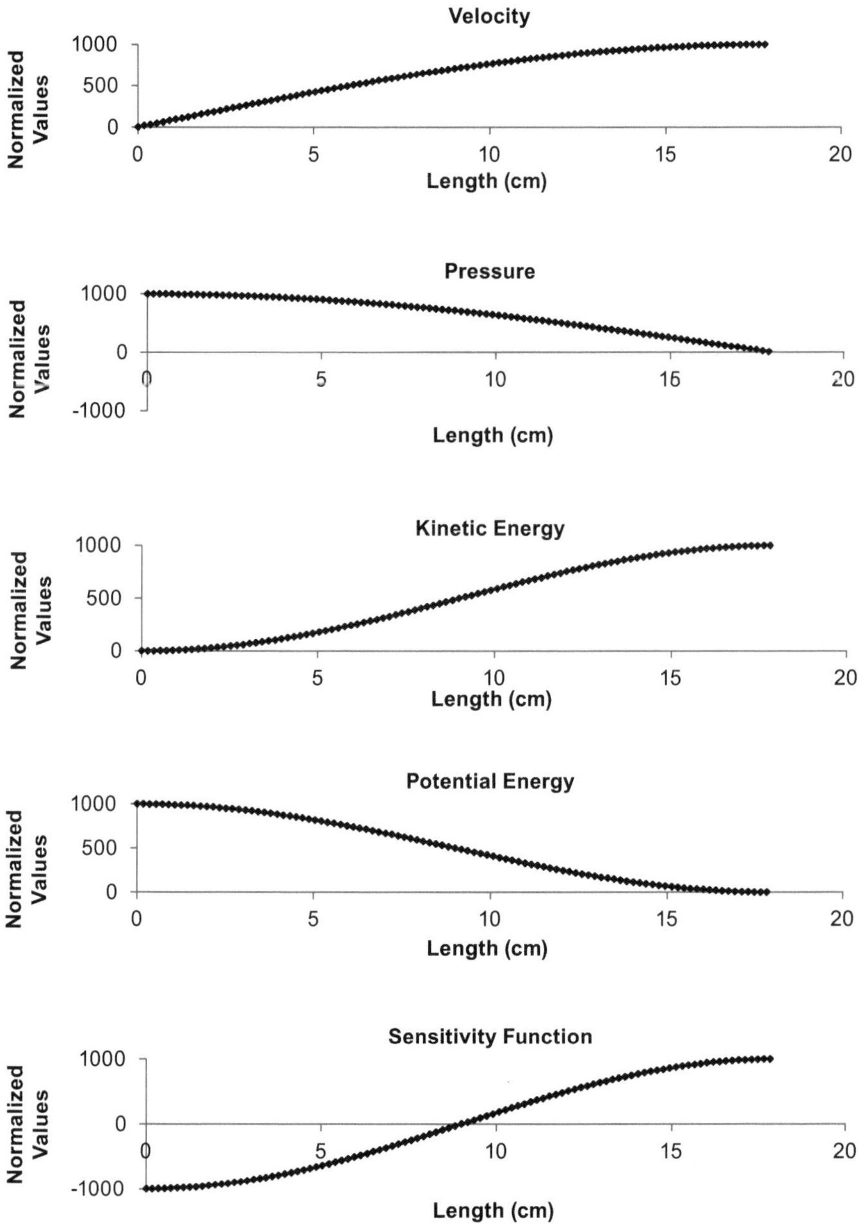

Fig. 2.2: Functions pertaining to the first resonance F_1 of the uniform tube, from top to bottom: spatial distribution of the velocity, pressure (scales normalized with respect to their individual absolute values), kinetic energy *KE*, potential energy *PE*, and the sensitivity function *SF*, all their scales normalized with respect to the maximum of the *TE*.

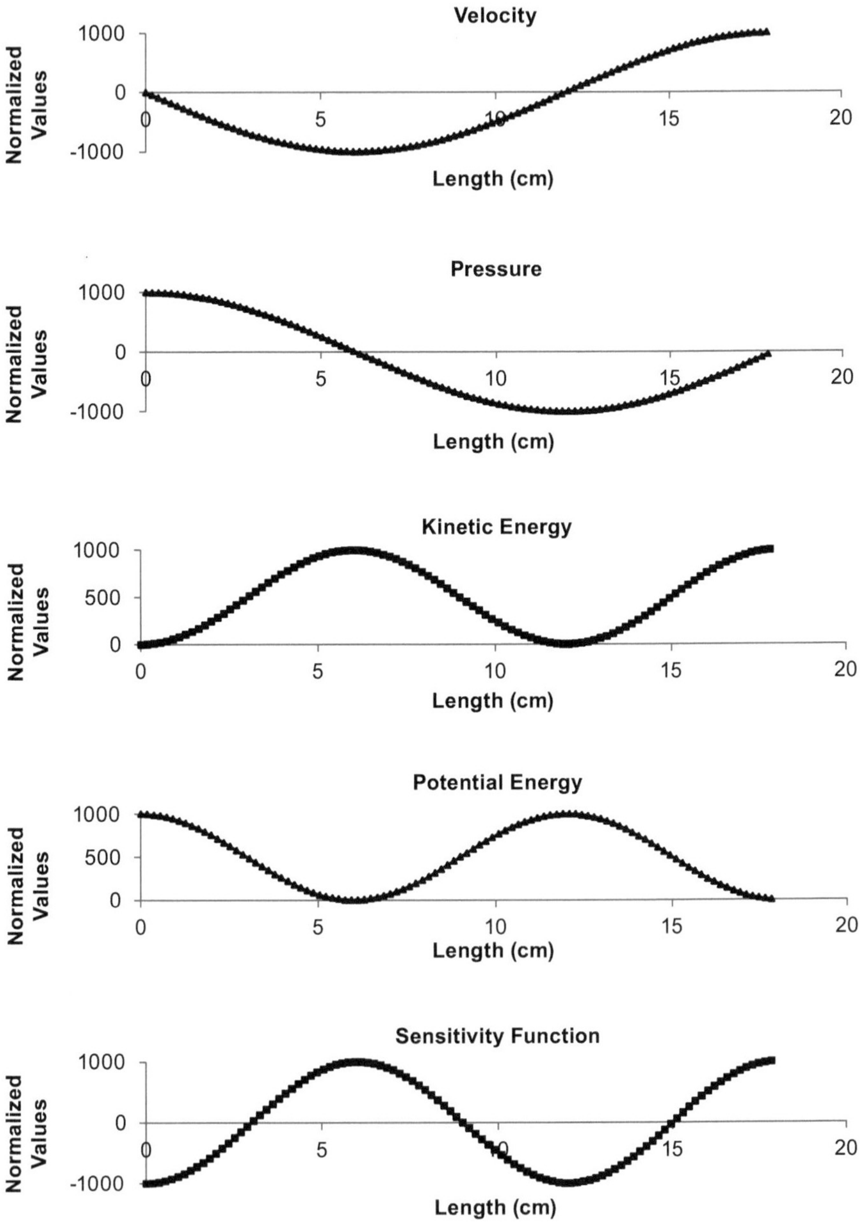

Fig. 2.3: Functions pertaining to the second resonance F_2 of the uniform tube, from top to bottom: spatial distribution of the velocity, pressure (scales normalized with respect to their individual absolute values), kinetic energy *KE*, potential energy *PE*, and the sensitivity function *SF*, all their scales normalized with respect to the maximum of the *TE*.

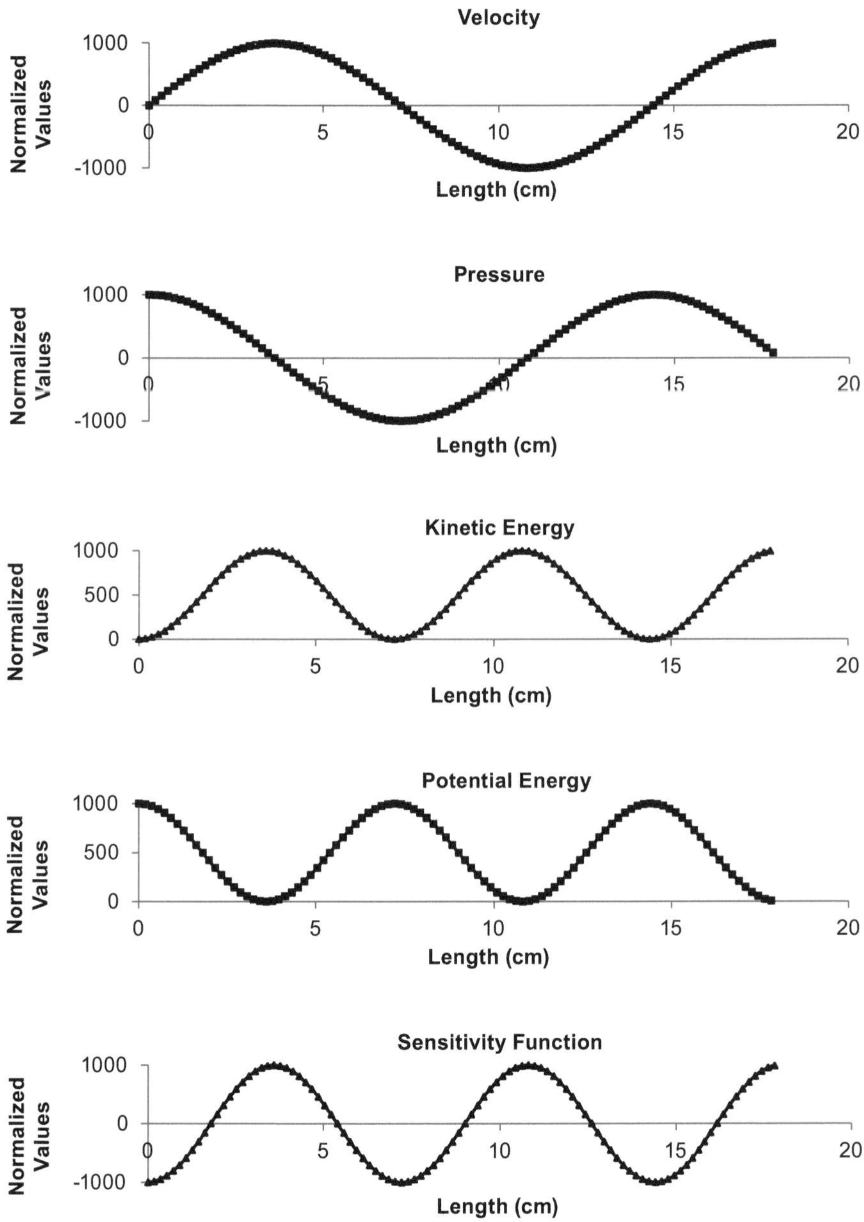

Fig. 2.4: Functions pertaining to the third resonance F_3 of the uniform tube, from top to bottom: spatial distribution of the velocity, pressure (scales normalized with respect to their individual absolute values), kinetic energy *KE*, potential energy *PE*, and the sensitivity function *SF*, all their scales normalized with respect to the maximum of the *TE* (after Fant and Pauli, 1974, Figure 1, p. 127, with permission by TMH KTH – Speech Communication and Technology Group).

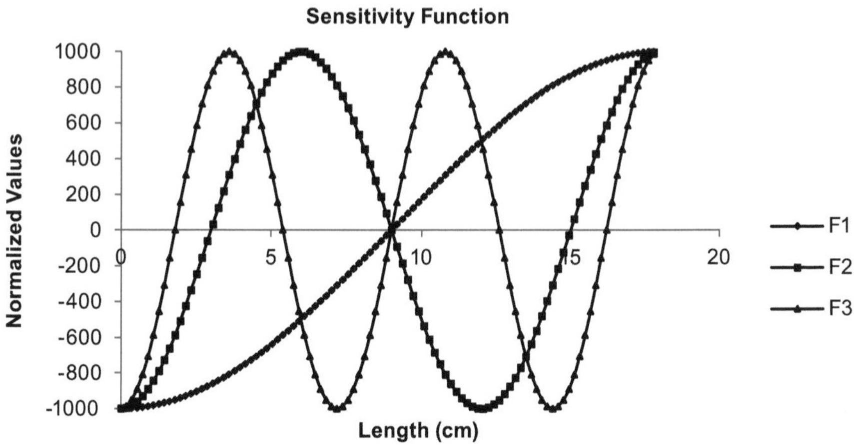

Fig. 2.5: Sensitivity functions of the three formants for a *uniform tube*. They are anti-symmetric and sinusoidal.

to the *difference* between the kinetic and the potential energy along the tube (the Lagrangian). The second deals with local area *longitudinal changes* and it is proportional to the *sum* of the two energies. Fant (1975b) showed that the sensitivity function for longitudinal perturbations is a measure of the relative dependence of the particular resonance mode on various parts of the area function. This is also the best definition for the "formant-to-cavity" affiliation (Fant, 1980). However, in the present book, sensitivity functions always refer to those affecting *transversal local area changes.*

2.5 Approximating the tube by N sections

When the continuous tube is approximated by N longitudinal sections, the area function at section $n < N$ is $A(n)$ and the sensitivity function becomes as follows (see Equations 2.6 and 2.7):

$$\begin{aligned}\frac{\Delta F_i}{F_i} &= \sum_{n=1}^{N} \frac{\overline{KE_i(n)} - \overline{PE_i(n)}}{TE_i(n)} \frac{\Delta A(n)}{A(n)} \\ &= \sum_{n=1}^{N} \overline{SF_i(n)} \frac{\Delta A(n)}{A(n)} \end{aligned} \tag{2.12}$$

where $\overline{SF_i}$ is the average sensitivity value over the tube section and it is a function of $A(n)$, i the formant number, and n the section number.

The kinetic and potential energies for each formant i are based on the average pressure $\overline{P(n)}$ and the average volume velocity $\overline{U(n)}$ computed for each section of the tube. They are calculated as (see Equations 2.2 and 2.3):

$$\begin{aligned} \overline{PE_i}(n) &= k_1 \overline{P}^2(n) A(n) \\ \overline{KE_i}(n) &= k_2 \overline{U}^2(n) / A(n) \\ \overline{SF_i(n)} &= \frac{\overline{KE_i(n)} - \overline{PE_i(n)}}{TE_i(n)} \end{aligned} \tag{2.13}$$

where k_1 and k_2 are constants.

The expressions in Equations 2.12 and 2.13 are informative and can provide help for the understanding of sensitivity functions in sound production phenomena. They show that, at any point along the tube, the kinetic energy density is proportional to the square of the flow velocity and inversely proportional to the area at that point. Conversely, the potential energy density is proportional to the area and to the square of the sound pressure. Consequently, when we reduce the area at a certain point in the tube, the kinetic energy is increased and the potential energy is decreased. This means that, at that point, the sensitivity function will tend to increase. Conversely, if we increase the area at a certain point, the sensitivity function will decrease.

2.6 Example on a relation between $A(x)$, $SF_i(x)$, and F_i

To demonstrate a pattern of change in the formant frequency F_i, change that corresponds to *small perturbation* of the area function $A(x)$ and that is proportional to the sensitivity function $SF_i(x)$, Figure 2.6 shows an example of the way F_1 is controlled (Carré, Lindblom, et al., 1995; Carré, 2004). The figure also displays the corresponding change of SF_i. Observing the figure leads to the following remarks:

- Starting from the neutral tube shape, the deformations of the area for increasing F_1 are the opposite of those for decreasing it.
- The sensitivity function remains sinusoidal.
- There is a synergetic effect, that is, simultaneous antisymmetric deformation of the front and back areas of the tube doubles the amount of change in F_1.
- Perturbing the area function in proportion to SF_i will keep the zero-crossing points of the area function and the sensitivity function unchanged.

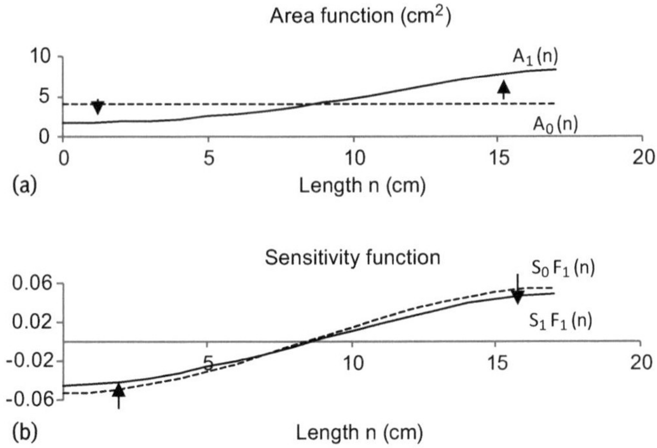

Fig. 2.6: For the uniform tube: Panel (a) increasing F_1 by deforming $A(x)$ by one step proportionally to SF_1; panel (b) corresponding changes in the sensitivity function.

2.7 An example of a non-uniform tube

For any tube shape, sensitivity functions also specify how well each resonant mode (i.e., formant) responds to a *local perturbation* of the area (expressed as a fixed, small, positive percentage) as a function of the position of the perturbation along the tube.

For the case of the uniform tube, any local positive perturbation ΔA in the front part of the tube leads to a positive first-formant change, while any local positive perturbation in the back part of the tube leads to a negative first-formant change because, as Figure 2.6 shows, the acoustic behavior of the closed-open uniform tube is *strictly* anti-symmetrical. However, when the tube is non-uniform (for example, a cone-shaped tube with area function linearly increasing from the input end as illustrated in Figure 2.7 a), the corresponding sensitivity function $S_0F_1(n)$ for the first formant (Figure 2.7 b) is no longer anti-symmetrical about the midpoint. The first formant change is maximum and positive at the aperture end of the tube, it is maximum and negative at about 5 cm, and it is zero at about 10 cm from the closed end. Thus, for this initial conical tube shape, the largest acoustic effects along the tube are obtained when deforming it at 5 and 18 cm, while the effect is nil when deforming it at 10 cm from the closed end, in an apparent correlation with the absolute magnitude of the sensitivity function. Deforming the initial area function in a way correlated with the sensitivity function of a given formant increases the frequency of that formant (see Carré, Lindblom, et al., 1995; Carré, 2004). Figure 2.7 a displays the area function obtained with a deformation leading to an efficient increase of F_1, and the corresponding new sensitivity function $S_1F_1(n)$. Subscripts 0 and 1 represent the initial and the post-deformation states, and are introduced in this example for clarity.

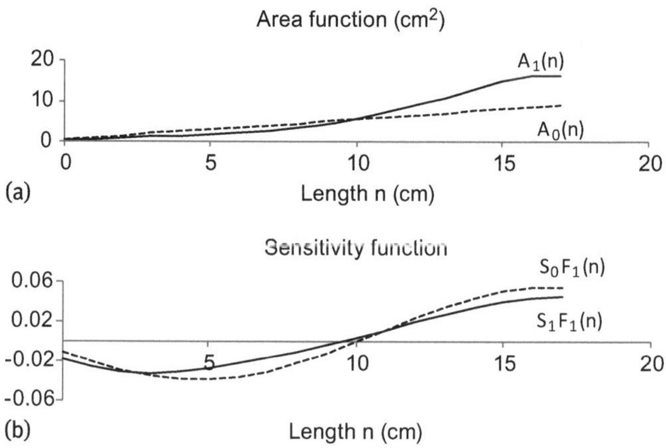

Fig. 2.7: Panel (a) Initial area function $A_0(n)$ (dotted line) of a non-uniform closed-open tube, and the new shape $A_1(n)$ (solid line) obtained by perturbing, in steps, the initial area function proportional to the sensitivity function $S_0F_1(n)$, in order to efficiently increase the first formant frequency. The subscript 0 indicates the area function's initial state and the subscript 1 its state after the first of the series of iterations. Panel (b) shows the sensitivity functions S_0F_1 and $S_1F_1(n)$ corresponding to the area functions in Panel (a).

Note that when the behavior of the initially conic-shape tube is compared to that of the initially uniform tube, some potentially important characteristics are lost. For example, we no longer have an anti-symmetry and therefore a *maximum synergetic* effect is not obtained. Furthermore, zero-crossings of the sensitivity function do not occur at the same point of the tube as they do when synergy is exploited in the initially uniform tube (as shown in Figure 2.6), and therefore the distinctive *sensitivity regions* along the tube are different for the two tube shapes.

Recall that the deformation must be small because the sensitivity function is only valid for a specific tube shape, although such small deformation leads to a maximum formant change. In other words, efficient deformation – maximum formant change for a minimum area deformation – is obtained when the shape of the area deformation function is similar to the shape of the sensitivity function.

2.8 Conclusions

In this chapter we explained and derived the concept of sensitivity functions, as they pertain to the tube in the focus of our interest. The length of the tube throughout this study is held constant at 18 cm, the area functions $A(n)$ contain 18 sections, each having a length of 1 cm, and the initial cross-sectional area of the uniform tube we consistently refer to has been always 4 cm^2. As will be mentioned in Chapters 3 and 4, the tube also incorporates the radiation impedance effect.

For the *uniform tube*, the sensitivity function equation (2.7) has led us to the following inferences:

1. SF_i changes sign at zero-crossing points, the region where it is positive will increase F_i with positive ΔA at that region and vice versa. The behavior is always anti-symmetrical with respect to the midpoint of the tube.
2. Perturbing the area function $A(x)$ (by applying a small deformation) in proportion to the SF_i of a certain F_i, will not change the other formants because changes in the front half of the tube *compensate* for changes in the back half.
3. Another aspect of the uniform tube is the *synergy* phenomenon, that is changing $A(x)$ in proportion to SF_i in one half of the tube and in the opposite way in the other half will double ΔF_i; thus $\Delta F_i/\Delta A(x)$ will be *maximum* and the process will be *efficient*. Because the process satisfies the requirements of control theory, this *maximally* efficient exploitation of the tube will also be *optimal*. These aspects therefore help in *controlling* the tube and can be used for modeling the process.
4. When $A(x)$ increases (towards a cavity) KE_i decreases and PE_i increases, consequently SF_i decreases, $\Delta A(x)/A(x)$ decreases, and therefore $\Delta F_i/F_i$ will also decrease. Inversely, when $A(x)$ decreases (towards a constriction), KE_i increases and PE_i decreases, consequently SF_i increases and $\Delta A(x)/A(x)$ increases, therefore $\Delta F_i/F_i$ will also increase, see Figure 2.6.
5. Successive perturbations of $\Delta A(x)$ in proportion to SF_i will not change the area function $A(x)$ at the zero crossing of the SF_i, because SF_i at that point is always zero, therefore $\Delta A(x)$ at that point is also zero. Zero crossing points become boundaries separating specific distinctive *regions* along the length of the tube.
6. Sensitivity functions for a closed/quasi-closed tube will be similar to those of the above-derived functions for the closed-open tube and the conclusions will also be similar, except that, instead of anti-symmetry, the behavior of the tube will be symmetric with respect to the midpoint.
7. To efficiently control any formant F_i, we should deform $A(n)$ in proportion to SF_i.

3 An efficient acoustic production system

The purpose of this chapter is to derive and explain the characteristics of the 18-cm long tube used for the generation of acoustic signals. This acoustic device can be used to generate sounds by applying excitation to the tube and by gradually changing its shape. But changing the shape of the tube leads to altering its resonant frequencies (i.e., its formants), that is, it results in acoustic variations. Figure 3.1 diagrammatically illustrates the main characteristics of the device. Using an iterative algorithm (Carré, 1995, 2004) guided solely by our criterion of maximum acoustic change for minimum deformation, the cross-sectional area of the tube is deformed step by step, as the simple flow chart in Figure 3.2 illustrates. The iterative process is terminated when the acoustic variation is no longer significant. The process results in a set of *efficient deformations* at *distinctive regions* producing *trajectories* that occupy the *largest possible formant space*. We want to emphasize that all along this procedure, the goal is acoustic and the task to reach it consists of deforming the shape of the tube. The iterative process is *inherently dynamic* in nature: it does not aim at reaching a succession of predefined static targets.

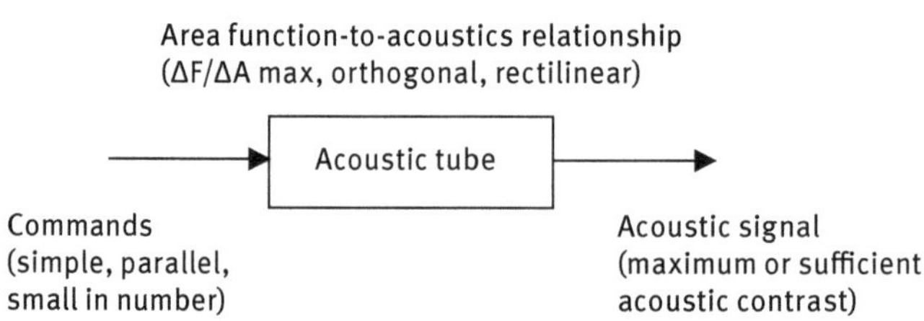

Fig. 3.1: Characteristics of the acoustic device used for optimal acoustic production. The commands are simple and small in number.

For this device to become an efficient and optimal system of acoustic sound production, several conditions must be met:

- To be efficient, the device must guarantee that the smallest possible overall deformation of the tube should lead to the largest possible acoustic variation (change in formants)
- Applying iteratively, step-by-step, a specific type of deformation to the tube should result in acoustic variations representing a monotonic trajectory in the formant space; the ensemble of all possible trajectories should outline boundaries of a maximum acoustic space. Exploitation of any sufficient subspace should also become possible.

DOI 10.1515/9781501502019-004

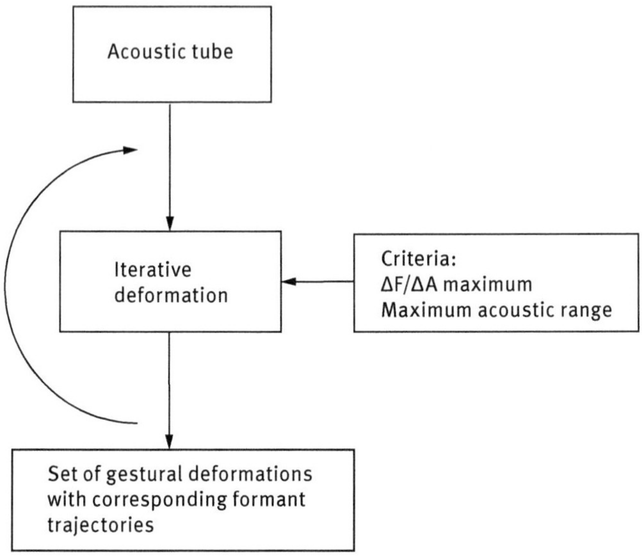

Fig. 3.2: Schematic representation of the iterative evolutionary process (ΔA: area function deformation, ΔF: corresponding formant frequency variation). Using an *iterative* algorithm, the cross-sectional area of the tube is deformed one small step at a time, guided by the sole criterion of obtaining *maximum acoustic variation* for *minimum deformation*. Results consist of a set of *efficient deformations* at *distinctive regions* producing longest acoustic trajectories.

- There should be a limited number of tube deformation types controlled by commands representing a set of codes that are simple, sparse, and fully used, whether separately or in parallel, by allowing different deformation types to be co-produced.
- Tube area deformations are by definition dynamic, such that the formant trajectories they produce are monotonic functions of time specified by their direction, velocity, and duration.
- The device must be operational even under unfavorable acoustic conditions – e.g., in the presence of interfering acoustic signals (low signal-to-noise ratio, SNR) or in excessive room reverberation.

Description of the dynamic aspects and optimal coding (including parallelism) of our system based on exploiting a simple acoustic tube device is deferred to Chapters 4, 5, and 6.

3.1 Algorithm for efficient change of tube shape

Starting from an initially uniform shape, how could one deform the area function of the acoustic tube, in order to obtain maximum acoustic variation with minimum shape deformation? To answer this question, we must examine the acoustic behavior of a tube in response to perturbations of its shape (i.e., the area function). As described in Chapter 2, the sensitivity function corresponds well to this requirement. When the area deformation is proportional to the sensitivity function for a specific resonance mode (i.e., formant), the smallest area function deformation (i.e., the least effort) will lead to maximum acoustic variation. The search for maximum acoustic variations means that the area function will not be deformed at places where deformations would have no acoustic effect.

Because the sensitivity function is only valid for small perturbations and depends on the shape of the tube, large formant changes cannot be simply extrapolated from results of small perturbations but have to be derived step-by-step according to the scheme in Figure 3.2. At each iterative step, the sensitivity function is recomputed and a new tube shape is obtained. Equation 3.1 represents the algorithm (Carré, 2004) used to deform the tube for mode (i.e., formant) i.

$$A(j+1,n) = A(j,n)[1 + k(n)q_i S_i(j,n)], \quad 0 \le j < J, 1 < n \le N \tag{3.1}$$

where:

i	is the serial number of the resonance mode, i.e., of the formant F_i;
N	is the total number of the sections of the 18-cm tube;
n	is the section number of the area function (between 1 and N);
j	is the iteration step number (between 0 and J);
$A(0,n)$	is the initial area function at section n of the tube;
$A(j,n)$	is the area function at section n of the tube at iteration j;
$S_i(j,n)$	is the sensitivity function of the formant i for $A(j,n)$;
q_i	is the weighting factor of the sensitivity function for the formant i;
$k(n)$	is the "deformability" coefficient of the tube (i.e., if $k(n) = 0$, the section n is not deformed).

The flow chart of the algorithm is shown in Figure 3.3 for the objective of increasing F_1. During the iterations, the area function $A(n)$ can vary between fixed limits $A_M(n)$ (maximum value) and $A_m(n)$ (minimum value). The size of a step of area variation depends on the weighting factors (Equation 3.1). This means that to reach $A_M(n)$ or $A_m(n)$, two factors can be used: J the total number of iterations, or q_i the weighting factors of the sensitivity functions. In the following experiments, J is fixed at 10 to describe the dynamics of the process with sufficient accuracy (to improve this accuracy, a larger number could be chosen but, because we are presenting the algorithm's results for sake of demonstration only, $J = 10$ is sufficient), then the weighting factors are adjusted to reach $A_M(n)$ or $A_m(n)$ at the end of the process.

Beyond the area range, the efficiency of the area deformation decreases and the iterative process can be stopped because it would be inefficient to continue it. Formant frequencies are calculated using the Badin and Fant (1984) algorithm.

The tube can also be deformed with the goal of simultaneously changing F_1, F_2, and F_3, that is, increasing or decreasing them depending on whether their respective weights q_1, q_2, and q_3 are positive or negative (Equation 3.2).

$$A(j+1,n) = A(j,n)[1 + k(n)(q_1 S_1(j,n) + q_2 S_2(j,n) + q_3 S_3(j,n))],$$
$$0 \le j < J; 0 < n \le N \tag{3.2}$$

In sum, a uniform tube shape can be efficiently deformed step-by-step, in order to obtain for each step the largest possible formant frequency change using the smallest area deformation. Departing from an initial uniform area function, the cumulative formant changes of these small deformations lead to acoustic trajectories embracing a formant frequency span that can grow as large as the adopted criterion [the range of $A(n)$] permits. This algorithmic tool thus represents an extension of Schroeder's (1967) and Fant's and Pauli's (1974) solutions for small tube deformations to large cross-sectional area deformations.

3.2 Computational results

Using the preceding algorithm, two questions will be addressed by computational methods. The first asks what the upper limits of the acoustic space, the F_1-F_2 plane, are. The second question asks how a uniform tube behaves when it is deformed such that its formant frequencies F_1, F_2, and F_3 increase or decrease independently from each other. Both questions will be addressed for the closed-open and the closed-closed tube types. Two sets of iterations are examined: one without the presence of a source cavity, having the tube constrained only by area limits, and one with the tube constrained by a source cavity and other more realistic factors borrowed from the human vocal tract. At the end of the chapter, a case of the algorithm with an initially non-uniform tube shape will be considered.

3.2.1 Exploring the F_1-F_2 acoustic plane

What is the largest acoustic space in which the tube designed for communication can operate? Using the algorithm (Equation 3.2) described in Section 3.1, we want

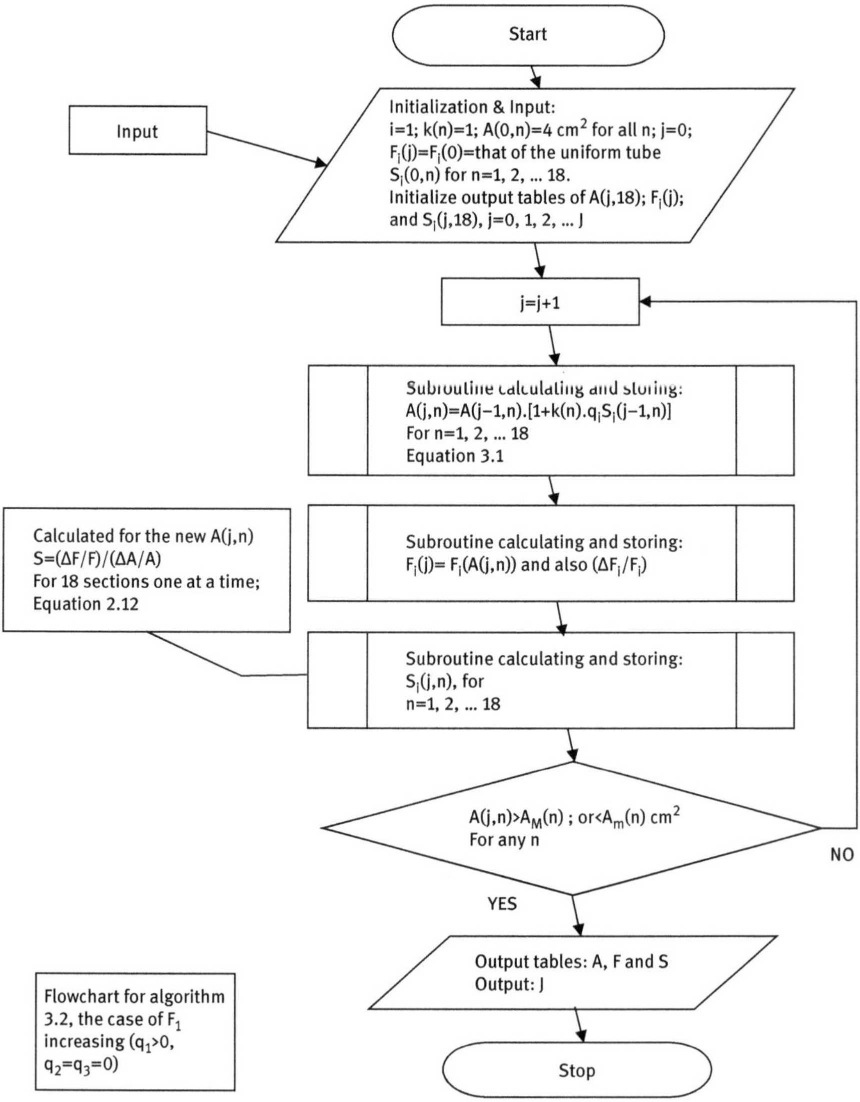

Fig. 3.3: Flow chart of the algorithm for the objective of increasing F_1.

to start from a uniform tube and find the maximum usable acoustic space (*lossy* case, see Subsection 3.2.3). In the initial experiment, the tube cross-sectional area is $4\,cm^2$, the possible range of the area function is between $A_m(n) = 0.5$ and $A_M(n) = 32\,cm^2$ (6 octaves around $4\,cm^2$), and the number of iterations is 10.

The iterative computation had a central starting point (the neutral position), defined by the resonances, that is the F_1, F_2, and F_3 values, of a uniform area section (i.e., cylindrical shape) tube. Using trigonometric functions for the weighting coefficients q_1 and q_2 (see Equation 3.2, $q_3 = 0$) with arguments ranging over a complete cycle, trajectories covering all directions of the F_1-F_2 space were obtained. Specifically, the coefficients were set to $q_1 = w \times \cos(\theta)$ and $q_2 = w \times \sin(\theta)$ for $0° < \theta < 360°$ in steps of 5° to compute 72 sets of deformations and their corresponding trajectories in the F_1-F_2 plane (Carré, 2004).

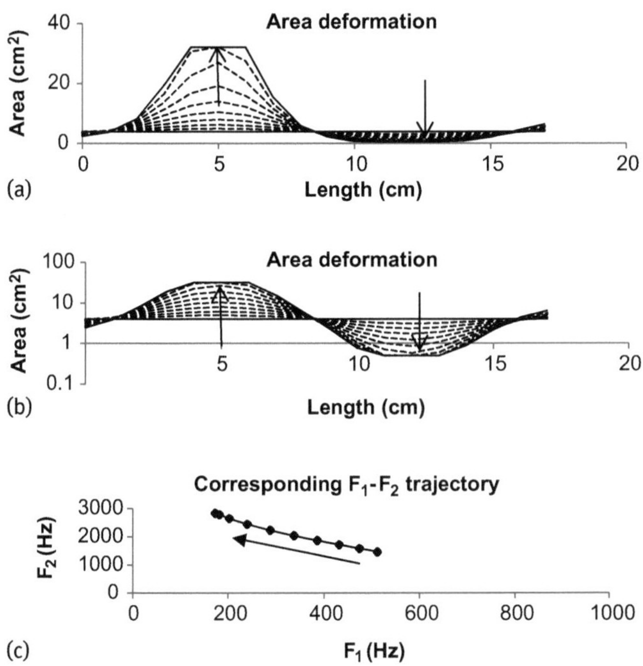

Fig. 3.4: Area deformation, linear scale (a), logarithmic scale (b), and (c) corresponding F_1-F_2 trajectory with $q_1 = -0.702 \times 3$ and $q_2 = +0.702 \times 3$ (see text). $A(0, n)$ is uniform and equal to $4\,cm^2$, the possible range of the area function is between $A_m(n) = 0.5$ and $A_M(n) = 32\,cm^2$ (6 octaves around $4\,cm^2$), the number of iterations is ten. The trajectory is one of the 72 trajectories computed to find the limits of the F_1-F_2 acoustic plane. The area deformation is anti-symmetric and the formant trajectory is rectilinear. At the end of the trajectory, the dots are closer meaning a decreasing of efficiency.

Figure 3.4 illustrates the results obtained for one example among the 72 computations (the weights are: $q_1 = -0.702 \times w$ and $q_2 = +0.702 \times w$ with $w = 3$): the area deformation is anti-symmetrical and the formant trajectory is rectilinear. The dots on the trajectory correspond to F_1-F_2 values obtained iteration by iteration; the distances between successive dots decrease at the end of the trajectory where the area deformation becomes less efficient due to getting close to the lower or the upper limits of the area (0.5 and 32 cm^2).

Figure 3.5 shows the 72 trajectories obtained from the neutral initial position in the F_1-F_2 plane. The corresponding F_1-F_3 trajectories are also represented. The triangle in the F_1-F_2 acoustic plane is limited, first, by the equation $F_1 = F_2$ and, second, by the presence of F_3. There is no limitation for increasing F_1 as long as

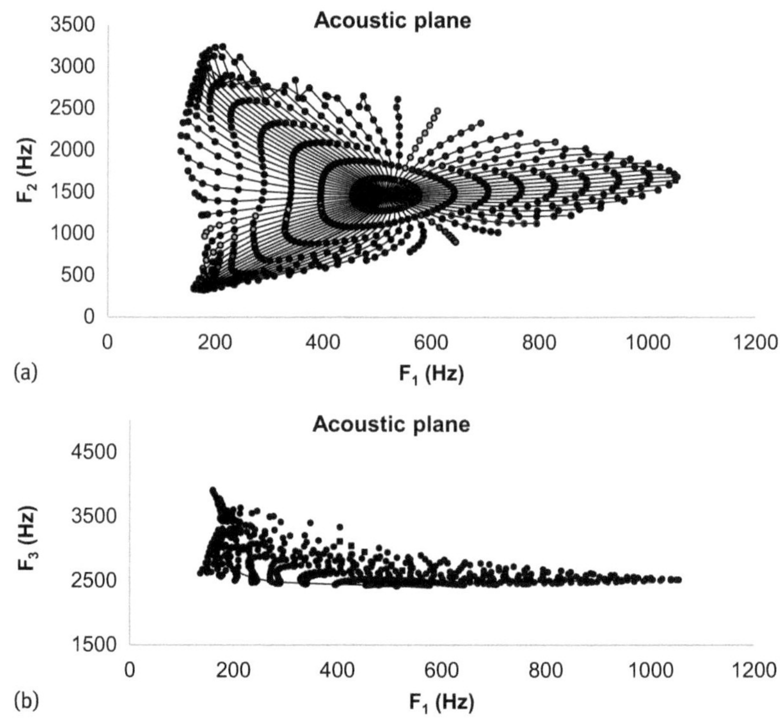

Fig. 3.5: Panel (a): Efficient usable F_1-F_2 acoustic space obtained from a closed-open uniform tube of 18 cm length; panel (b): corresponding F_1-F_3 acoustic space. $A(0, n)$ is uniform and equal to 4 cm^2, the possible range of the area function is between $A_m(n) = 0.5$ and $A_M(n) = 32$ cm^2 (6 octaves around 4 cm^2), $q_1 = \cos(\theta) \times 3$ and $q_2 = \sin(\theta) \times 3$, the number of iterations is 10.

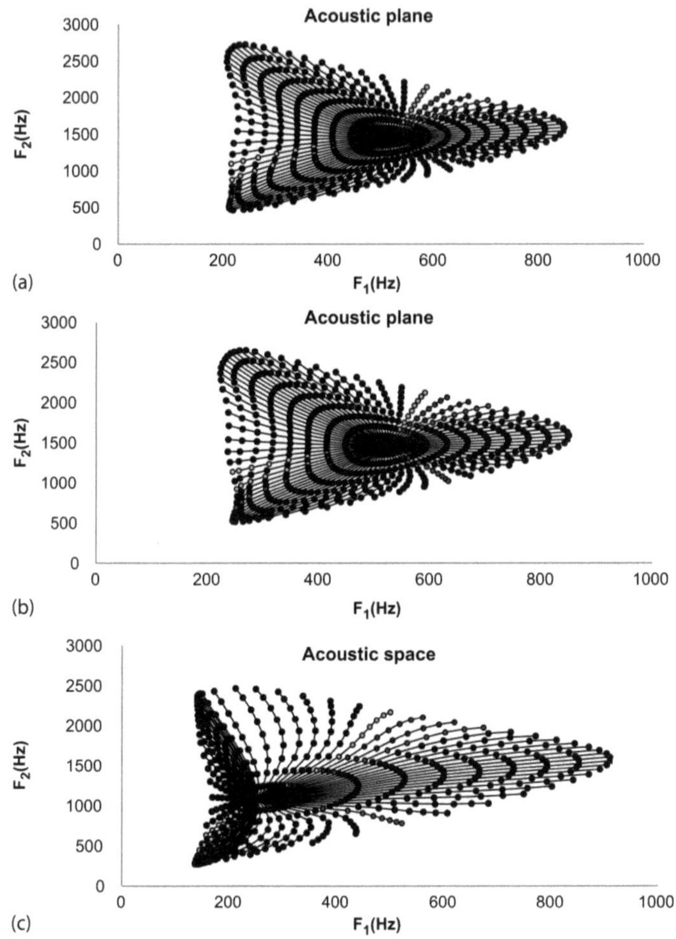

Fig. 3.6: (a) F_1-F_2 acoustic plane obtained from a uniform tube cross-sectional area of 4 cm^2, possible range of the area function is between $A_m(n) = 0.5$ and $A_M(n) = 32\,\text{cm}^2$, $q_1 = \cos(\theta) \times 1.8$ and $q_2 = \sin(\theta) \times 1.8$, number of iterations is ten; (b) from a uniform tube cross-sectional area of 2.82 cm^2, possible range of the area function is between $A_m(n) = 0.5$ and $A_M(n) = 16\,\text{cm}^2$, $q_1 = \cos(\theta) \times 1.8$ and $q_2 = \sin(\theta) \times 1.8$, number of iterations is ten; (c) from a closed-closed uniform tube of 4 cm^2, possible range of the area function is between $A_m(n) = 0.5$ and $A_M(n) = 32\,\text{cm}^2$, $q_1 = \cos(\theta) \times 5$ and $q_2 = \sin(\theta) \times 5$, number of iterations is 10.

it stays smaller than F_2. Compared to the triangle in the F_1-F_2 plane, the area of the triangle in the F_1-F_3 plane is small.

In order to gain a better understanding of the limits of the triangle in the acoustic plane, three complementary experiments were conducted, with its results shown in Figure 3.6. In order for F_2 to always stay below F_3, the space to be explored must be reduced.

In Figure 3.6 a, the uniform tube cross-sectional area is 4 cm^2, the possible range of the area function is between $A_m(n) = 0.5$ and $A_M(n) = 32$ cm^2, the weighting functions are $q_1 = \cos(\theta) \times 1.8$ and $q_2 = \sin(\theta) \times 1.8$ (i.e., smaller than in Figure 3.5, in order to keep the space to be explored smaller), and the number of iterations is 10. The decrease of the weighting functions leads to a decrease of the distance between dots.

In Figure 3.6 b, the uniform tube cross-sectional area is 2.82 cm^2, the possible range of the area function is between $A_m(n) = 0.5$ and $A_M(n) = 16$ cm^2 [5 octaves around 2.82 cm^2, instead of 6 octaves to study the effects of the range of $A(n)$], the weighting functions are $q_1 = \cos(\theta) \times 1.8$ and $q_2 = \sin(\theta) \times 1.8$, and the number of iterations is 10. The results are more or less the same – more uniform distance between the dots, better linearity.

In Figure 3.6 c, the tube is a closed/quasi-closed uniform tube. Its cross-sectional area is 4 cm^2, the possible range of the area function is between $A_m(n) =$ 0.5 and $A_M(n) = 32$ cm^2, the weighting functions are $q_1 = \cos(\theta) \times 5$ and $q_2 = \sin(\theta) \times 5$, and the number of iterations is 10. The outline of the trajectories obtained closely resembles the quasi-triangle obtained for the closed-open tube. It seems therefore that the type of tube has no influence on the iterative process launched to find the size of the usable acoustic space.

In these three experiments, the outline of the last points of the 72 trajectories resembles a triangle. The limits are due to the line $F_1 = F_2$ and to the F_1-F_3 space (because F_2 cannot cross F_3). In addition, the limits for increasing F_1 alone are only due to the range of $A(n)$ (in octaves). Figure 3.7 shows the area deformations (between 0.25 and 64 cm^2 – 8 octaves) and the corresponding F_1-F_2 trajectory. The weighting functions are: $q_1 = 3$ and $q_2 = 0$, the number of iterations is 15. Compared with the preceding experiments, much larger F_1 variations are obtained.

We performed several additional experiments similar to those mentioned above, and we came to the conclusion that when $A(n)$ is limited to the range between $A_m(n) = 0.5$ and $A_M(n) = 16$ or 32 cm^2, with the weighting functions $q_1 = \cos(\theta) \times 1.8$ and $q_2 = \sin(\theta) \times 1.8$ performing 10 iterations leads to the best compromise between “maximum acoustic contrast” and “global” efficiency.

In the F_1-F_2 plane the distances between successive dots decrease as they move away from the center. This indicates that the sensitivity of a given deformation decreases, that is, it becomes less and less efficient to produce an acoustic change as it moves farther and farther away from the initial neutral state. This means that the efficiency is maximum at the center corresponding to the uniform tube shape (as discussed in Chapter 2). Inside a *maximum* acoustic contrast plane (that will become the vowel triangle as will be discussed in Chapters 5, 6, and 7)

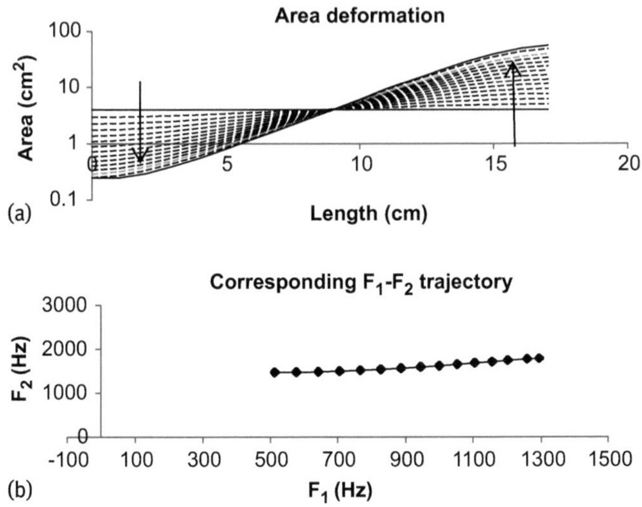

Fig. 3.7: Area deformation, (a) logarithmic scale and (b) corresponding F_1-F_2 trajectory. $A(0, n)$ is uniform and equal to 4 cm^2, the possible range of the area function is between $A_m(n) = 0.25$ and $A_M(n) = 64\ \text{cm}^2$ (8 octaves around 4 cm^2), the weighting functions are: $q_1 = 3$ and $q_2 = 0$, the number of iterations is 15.

the *sufficient* acoustic contrast plane must be centered around the neutral. This also means that the skirts of the maximum contrast plane (vowel triangle) are rarely or never used, suggesting that precise knowledge of the limits is not required. Figures 3.6 and 3.7 show aspects of the non-linear relation between area function change and formant frequency change – a linear relation would have resulted in circles (its radius being a function of the scale) around the initial central point. As the distance from the center increases, the diameter of the circle non-linearly decreases. On the other hand, the trajectories are approximately linear. Similar linearity was observed by Gunnilstam (1974, p. 91).

3.2.2 Behavior of an initially uniform tube constrained only by area limits

Using the preceding algorithm (Equation 3.1), our objectives were (1) first efficiently increase and then decrease F_1 (by obtaining maximum acoustic variation for minimum deformation all along the tube, Figure 3.8), then to do the same for F_2 (Figure 3.10) and F_3 (Figure 3.12); (2) let the algorithm proceed up to the end of this process (constrained by the limitations of $0.5 < A(n) < 32\ \text{cm}^2$ and $q_j = +1.8$ or -1.8).

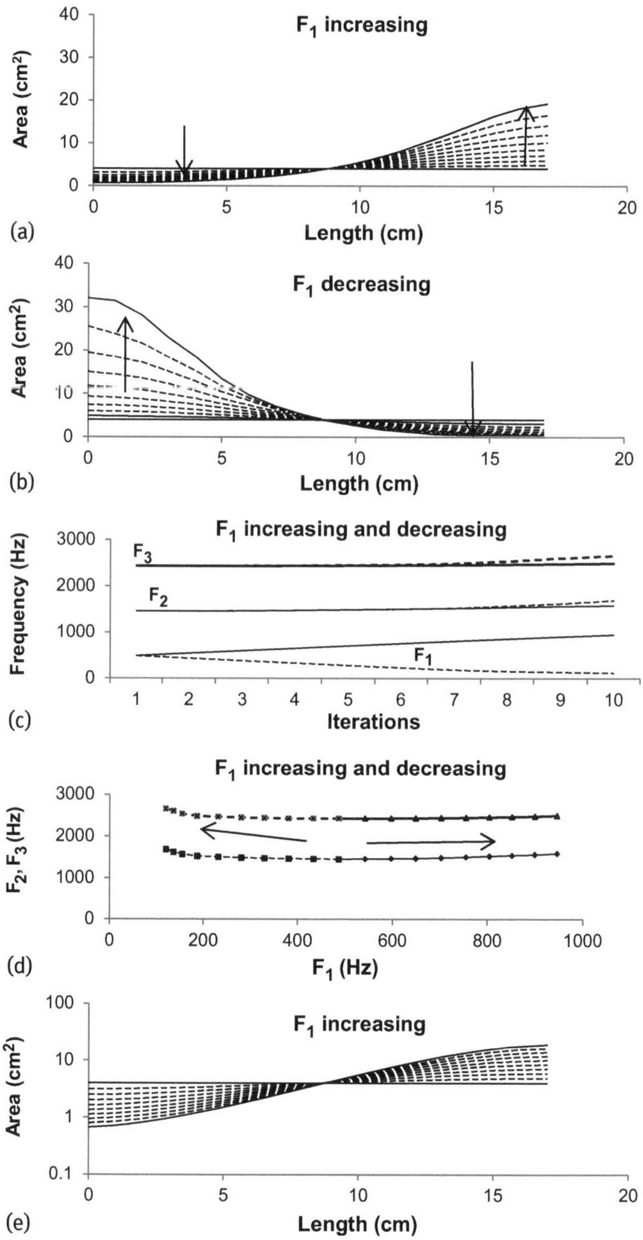

Fig. 3.8: Effects of efficient, minimum deformations over ten iterations of a closed-open initial tube shape (uniform tube) directed to increase F_1 [$q_1 = 1.8$, panel (a)] or decrease F_1 [$q_1 = -1.8$, panel (b)]. The corresponding formant variations as a function of iterations [panel (c)] and F_1-F_2 and F_1-F_3 trajectories [panel (d)]; solid lines illustrate increasing, and broken lines decreasing formant frequencies. F_1 frequency increasing as in panel (a) except the area function ordinate is on a logarithmic scale [panel (e)].

These objectives were implemented for a lossless closed-open tube (with the source situated at the closed end and the output side at the open end) and for a lossless closed-(quasi-) closed tube (i.e., a tube having a closed input end and a very small 0.1-cm^2 opening at the output end). No source cavity was included. The initial state (i.e., iteration number 0) of the area function $A(0, n)$ at each point n of the tube length was the cylindrical tube having an 18-cm length and a uniform cross-sectional area of $4\,cm^2$. The areas varied between 0.5 and $32\,cm^2$ (corresponding to an area range of 6 octaves) and $k(n)$, the "deformability coefficient," was set to 1.0 all along the tube.

3.2.2.1 Closed-open tube

Trajectories obtained by the iterative algorithm using minimum deformation to change F_1 and F_2 are shown in Figures 3.8 and 3.10 (F_3 is also represented in the figures). The results illustrated in the figures lead to some interesting observations.

For F_1, maximum and minimum values are obtained when anti-symmetrical deformations are applied to two regions of the tube (one in the front and one in the back), with a back constriction paired with a front cavity enlargement, and vice versa (Figure 3.8 a, b). It can be seen that the deformation is acoustically most efficient for F_1 between approximately 200 and 800 Hz and rapidly reaches asymptotic limits above and below that range. Within this range, for a logarithmic ordinate scale the deformed area functions can be approximated by a half-cosine function between π and 2π (in other words, the deformation is anti-symmetric, see Figure 3.8 e). For any given deformation the area functions will differ from the corresponding sensitivity functions only by a scalar in the opposite phase. Also, the sensitivity functions will be sinusoidal and will have the same zero-crossings as the area deformations (Figure 3.9). It can also be seen that, except for the lowest F_1 values, F_2 stays constant when F_1 changes (Figure 3.8 c, d). In fact, Schroeder (1967) already showed that when the shape of a uniform tube is slightly deformed

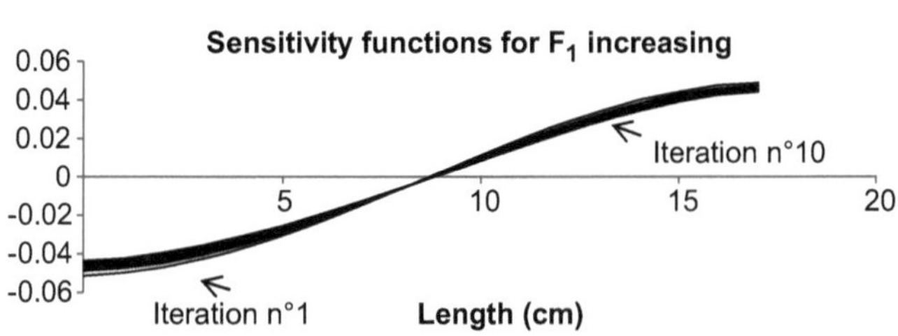

Fig. 3.9: Successive sensitivity functions for F_1 increasing corresponding to Figure 3.8 a.

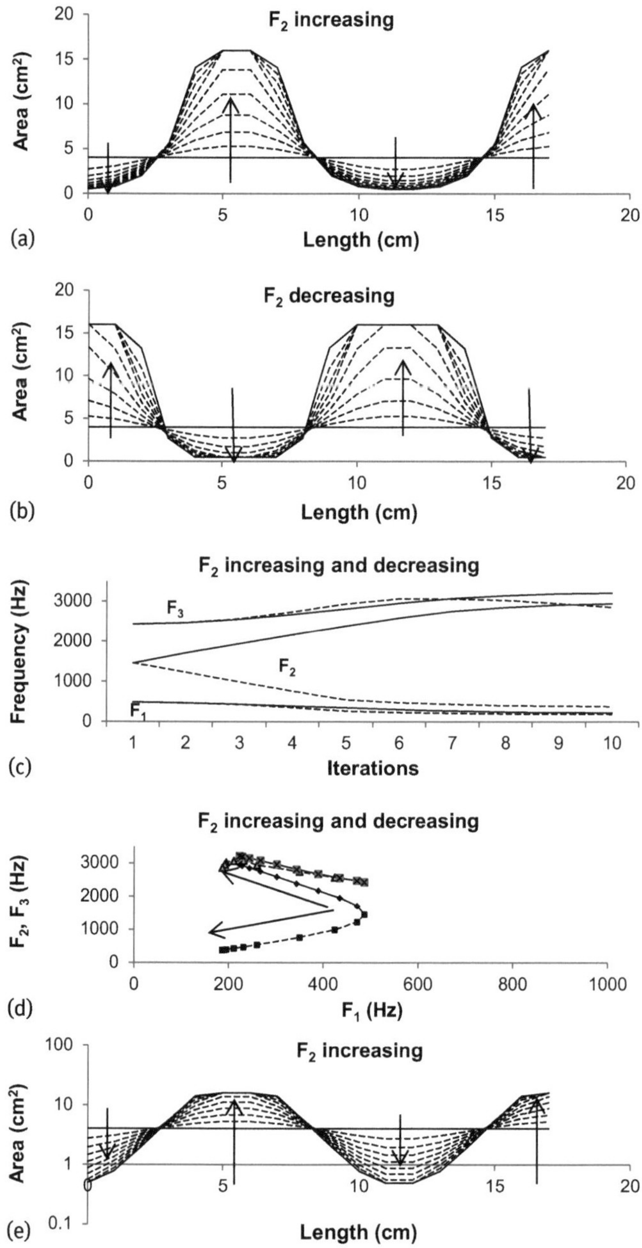

Fig. 3.10: Minimum area function deformations [panels (a) and (b)] and corresponding maximum formant frequency changes [panel (c)] and formant trajectories [panel (d)] resulting from deformations of an initially uniform shape closed-open tube by the algorithm directed either to increase (panels (a) and (c), $q_2 = 1.8$) or decrease (panels (b) and (d), $q_2 = -1.8$) F_2 over ten iterations; formant increase, logarithmic scale [panel (e)].

using an anti-symmetrical 0.5- (or 1.5-, 2.5-, etc.) cycle sinusoidal function, the resulting frequency variations are limited to a single formant with all other formant frequencies, for all practical purposes, remaining constant. Our results affirm Schroeder's observations for F_1 and extend them by establishing their range of validity.

For formant F_2, the algorithm yields four deformation regions (Figure 3.10) of unequal region lengths. The lengths of the regions are proportional to the sequence 1-2-2-1 and the area function along the tube can be approximated by a 1.5-cycle sinusoid. The anti-symmetry of the area function, i.e., the π phase shift between the two ends of the tube, is as predicted by the structure of closed-open tubes that is anti-symmetrical by definition.

The F_1 data from Figure 3.8 a, b can be approximated by a *linear* function of the diameter of the front cavity (section 18) of the tube (Figure 3.11 a), while the F_2 data taken from Figure 3.10 a, b adequately fit a *logarithmic* function of the diameter of the front constriction (section 13) of the tube (Figure 3.11 b). It is interesting to note that these two scales make the diameter of the tube produce a frequency relation much like the ERB scale (Chapter 8) in audition (i.e., quasi-linear at low and quasi-logarithmic at high frequencies), suggesting that the perceptual system formed by the ear may be optimally matched to the production system that a simple tube represents.

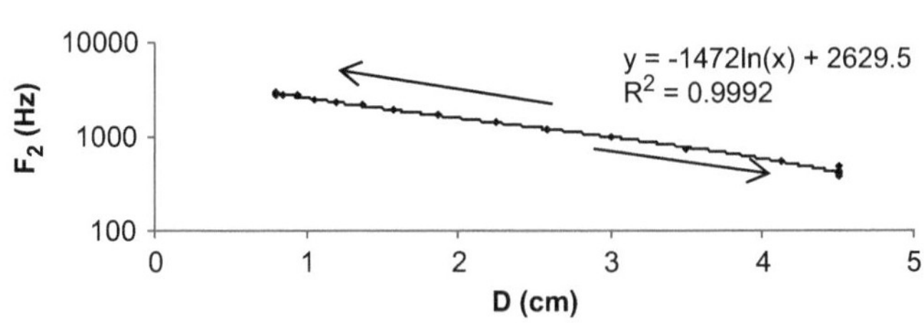

Fig. 3.11: Formant variations as a function of the diameter of the constriction. (a) F_1 variations as a function of the diameter of the opening (section 18, data from Figure 3.8 a, b); (b) F_2 variations as a function of the diameter of the constriction (section 13, data from Figure 3.10 a, c).

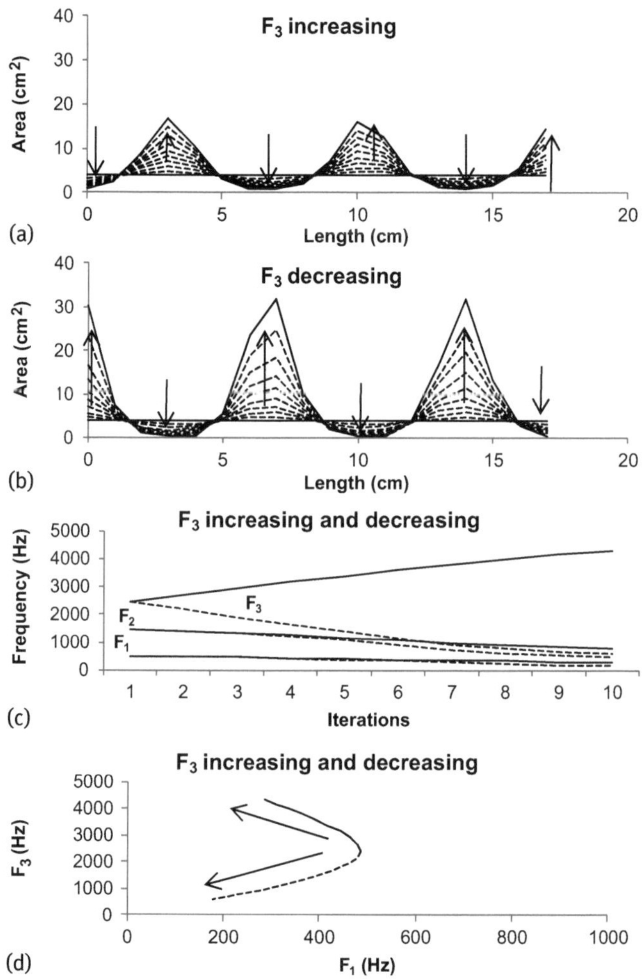

Fig. 3.12: Panel(a): Deformations of a closed-open initial tube shape (uniform tube) with the goal of increasing F_3 ($q_3 = 1.8$). Panel (b): Same but with goal of decreasing F_3 ($q_3 = -1.8$). Panel (c): Formant frequency changes corresponding to the area functions in Panels (a) and (b). Panel (d): Formant trajectories corresponding to the same area functions in Panels (a) and (b) (solid lines: increasing, dashed lines: decreasing).

When manipulating F_3, the algorithm yields six regions (Figure 3.12). Again the region lengths are unequal and proportional to a 1-2-2-2-2-1 sequence of region lengths, approximated by a 2.5-cycle sinusoid.

3.2.2.2 Closed-closed tube

After showing the behavior of the closed-open acoustic tube, the closed-closed tube has also to be examined. However, investigating the acoustics of this tube type has an inherent problem. To measure acoustic output, there has to be measurable output energy. But, in order to have any output energy, the tube could not

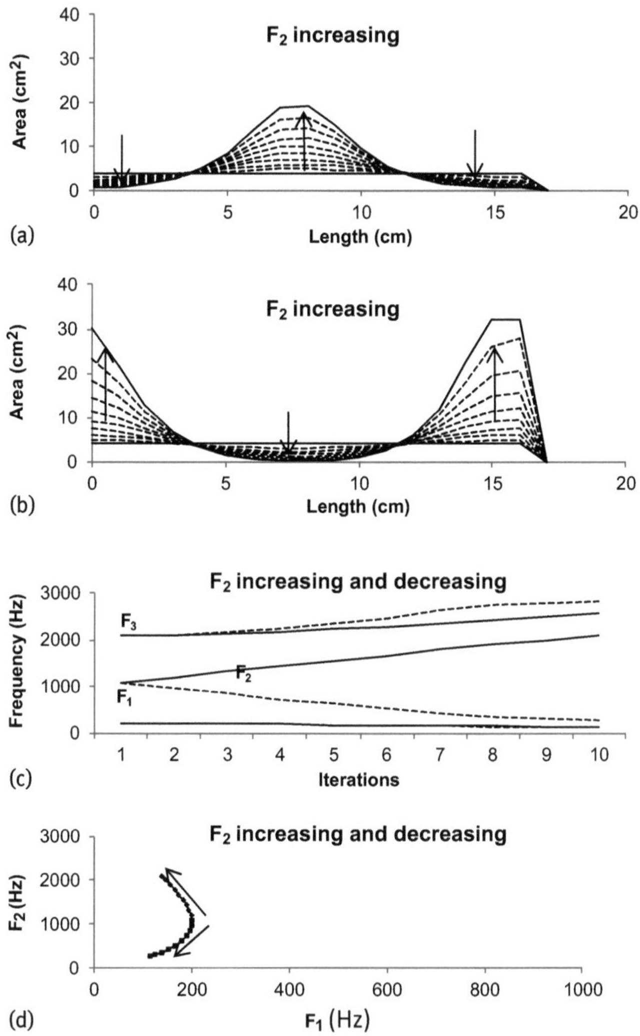

Fig. 3.13: Deformations of an initial closed-closed tube shape (uniform tube) in order to increase or decrease F_2 (a and b) with the corresponding formant variations (c) and efficient formant trajectories (d) over ten iterations.

be closed, that is the area of the output section could not be zero. As a compromise, the output area in the algorithm had a non-zero, 0.1 cm^2 value throughout, thereby creating a *practically* closed-closed tube. Because the equations shown in Section 3.1 hold regardless of whether either end of the tube is open or closed, the algorithms used for the closed-open tube in the previous subsection were adopted also for the closed-closed tube type. Obviously, for a completely closed end F_1 would be 0 Hz at the output at an energy level of 0 dyne. In the closed-quasi-closed tube with an opening area of 0.1 cm^2, deforming the tube to manipulate F_1 results in small F_1 frequency changes around 200 Hz. When starting from a uniform initial state, the algorithm makes the behavior of F_2 symmetrical (just as the closed-closed tube configuration itself is symmetrical), and the resulting area function changes can be approximated by one complete sinusoidal cycle (see Fig-

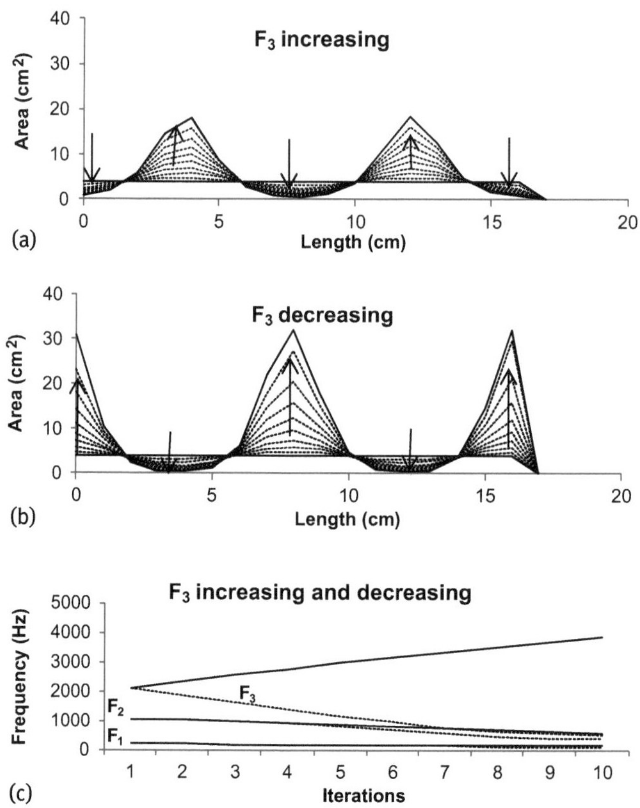

Fig. 3.14: Deformations of an initial closed-closed tube shape (uniform tube) with increasing or decreasing F_3 (a and b), with the corresponding formant variations (c).

ure 3.13). With F_2 added, three regions and, with F_3 also taken into account, five regions are obtained (Figure 3.14). Although the closed-closed tube increases the acoustic contrast for low F_1 and F_2 frequencies better than the closed-open one, but due to its output amplitude being much lower than that of the closed-open tube, it is definitely less well adapted for acoustic communication.

3.2.3 Toward a more realistic sound-producing tube

The system presented so far represents a simple acoustic tube; no a priori knowledge of the human vocal tract was assumed to build it. However, even without wanting to have it become an analog of the human speech production system, we were wondering if its functionality could improve by incorporating selected characteristics of the vocal tract. The characteristics chosen continued the path anchored in physics followed so far; their adoption would merely make the tube more realistic. Along these lines, the acoustic tube will be modified by four characteristics: (1) limit the area range to between 0.5 and 16 cm^2, (2) adding a fixed source cavity, (3) adding radiation impedance, and (4) adding wall vibration. Our results for more ample area ranges show that an area range between 0.5 to 10–16 cm^2 (about five octaves) is *efficient*, just as Fant has suggested (1960).

The primary *sound source* in the human speech production system consists of the vocal folds: an organ capable of generating a train of broadband, mostly periodic glottal pulses that excite the resonance cavities of the vocal tract. The vocal folds are enclosed in a fixed-size laryngeal cavity having a cylindrical shape: it has a length of about 2 cm and a uniform area of 2.5 cm^2 (Fant, 1960). Consistent with this fact, a cavity of such size has been imposed on the first two 1-cm sections of the uniform tube, although in this chapter no excitation source is used. The formant frequencies are calculated based on the transfer function of the tube with this source cavity added (Badin and Fant, 1984).

The *radiation effect* has two components: one inductive and one resistive. The inductive part affects the total effective length of the tube, L_e, equal to the physical length L_p plus an added length that is a function of the radiation inductance effect L_r related to the lip opening's area A_l as the formula $L_r = 0.8\sqrt{A_l/\pi}$ shows (Fant, 1975a). However, in human speech production L_e is nearly constant despite the lengthening of the protrusion associated with lip rounding. For example, when $A_l = 16\,cm^2$ L_r is 1.8 cm; for $A_l = 0.5\,cm^2$ it is 0.3 cm. "Lip rounding" corresponding to $A_l = 0.5\,cm^2$ is generally associated with a protrusion of about 2 cm. Therefore, as a first approximation, protrusion length (0 or 2 cm) counteracts – and just about cancels – the radiation effect (1.8 or 0.3 cm). Since, according to Fant, contributions of the losses due to the resistive portion of the radiation (as well as to heat and viscosity) are negligible, only the inductive portion of the radiation was taken into consideration in our modified acoustic tube.

The *wall vibration effect* was defined by Fant (1960) as $F_n^2 = F_{ni}^2 + F_w^2$, where F_w is the resonance frequency of the vocal tract closed at both ends, F_{ni} is the n-th resonance frequency of the hard-walled system, and F_n is the resulting resonance frequency (Mrayati and Carré, 1976). The wall vibration effect is non-negligible and has been therefore included into the realistic version of the tube.

3.2.3.1 The realistic closed-open tube

Effects of adding the four factors discussed above are shown first on the F_1-F_2 acoustic plane with the new acoustic triangle (Figure 3.15). Compared with the acoustic triangle obtained without source cavity (Figure 3.6 a), a small reduction is observed.

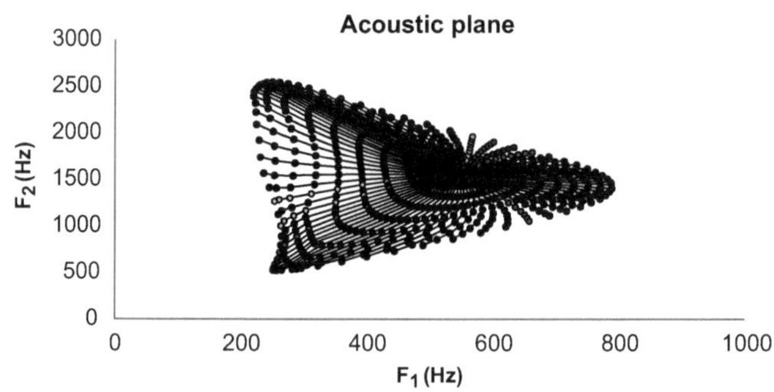

Fig. 3.15: F_1-F_2 plane for a uniform tube with a source cavity. The possible range of the area function is between $A_m(n) = 0.5$ and $A_M(n) = 16\ \text{cm}^2$, $q_1 = \cos(\theta) \times 1.8$ and $q_2 = \sin(\theta) \times 1.8$, number of iterations is ten.

Adding the source cavity and the losses for increasing or decreasing F_1, leads to the results shown in Figures 3.16 and 3.17 for increasing and decreasing F_2, respectively. When comparing results of the original tube (Figures 3.7 and 3.10 in Subsection 3.2.2) to those of the realistic tube (Figures 3.16 and 3.17), one may notice that (1) the range of acoustic variations exhibit only a slight decrement for the realistic tube, and (2) that although the region boundaries have moved, they did so only slightly. (The reader may recall that around region boundaries the sensitivity functions are close to zero; thus a small displacement of the boundaries results in a minuscule acoustic effect). Notice also that limiting the cross-section area variation between 0.5 and 16 cm^2 introduces saturation effects at the skirts of the acoustic triangle.

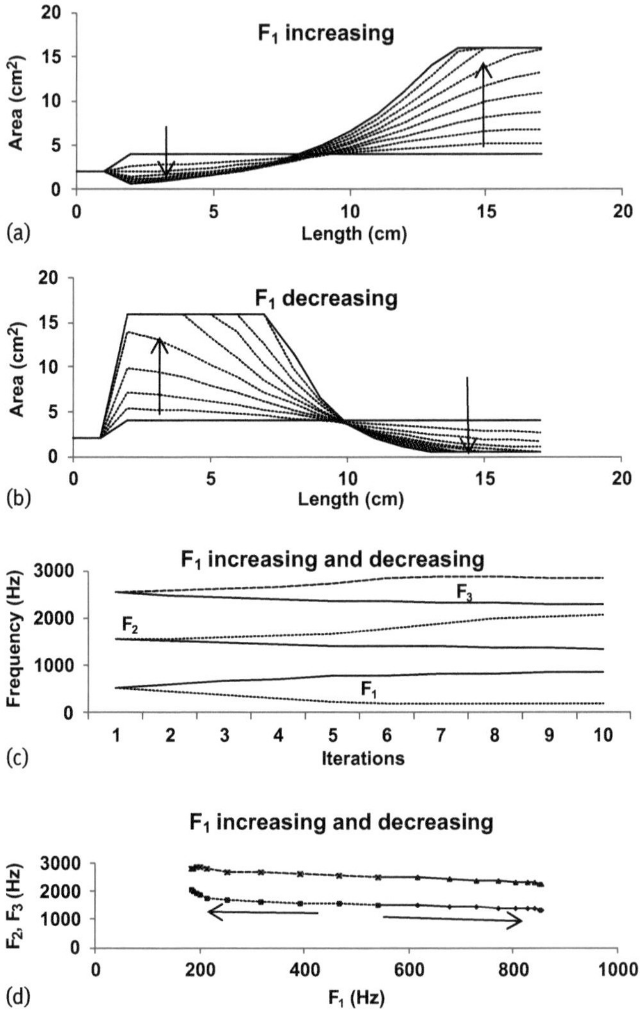

Fig. 3.16: Deformations of an initial closed-open tube shape (uniform tube with source cavity) to increase or decrease F_1 (a and b) with the corresponding formant variations (c and d) during ten iterations.

3.2.3.2 Closed-closed tube

In the case of the closed-closed tube, inclusion of source cavity and wall vibration effects also leads to small acoustic variations (not shown here). For example, for the cylindrical (i.e., uniform area) tube configuration F_1 is now 250 instead of 200 Hz (this value is due to the "lip opening" fixed at $0.1\,\text{cm}^2$ and the wall vi-

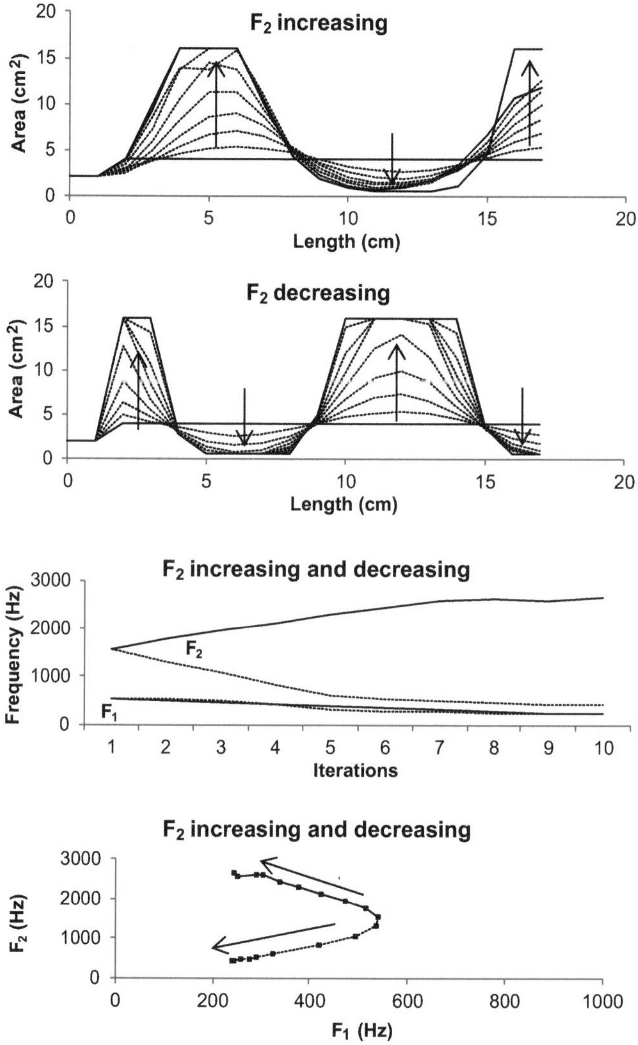

Fig. 3.17: Deformations of an initial closed-open tube shape (uniform tube with source cavity) aimed at increasing or decreasing F_2 (a and b) with the corresponding formant variations (c and d) over ten iterations.

bration effect with $F_W = 180\,\text{Hz}$). In other words, all observations made in the preceding subsections for the original tube remain acceptable also for the realistic tube.

3.2.3.3 Passage from a closed-open to a closed-closed tube

We have shown in the above subsections two different tube configurations, the closed-open and the closed-closed one. However, since humans have only one tube, we will have to assume that its configuration could change from the first to the second or the other way around. Therefore, we would like to examine the passage between the two configurations, in particular, from the closed-open to the closed-closed one. For the passage to occur, the simplest way is to gradually close the front opening. As, for the uniform closed-open tube, the area at the front opening is the same as everywhere else along the tube (except for the source cavity), gradually shrinking that area to the extreme, i.e., to the point at which it reaches zero, the tube eventually becomes fully closed-closed. To investigate the course of this transition, the preceding algorithm was used with various fixed openings at the tube's aperture opposite to the source, ranging from 16 to 0.1 cm^2 in logarithmic steps, so that intermediate situations between open and closed tube aperture could be evaluated.

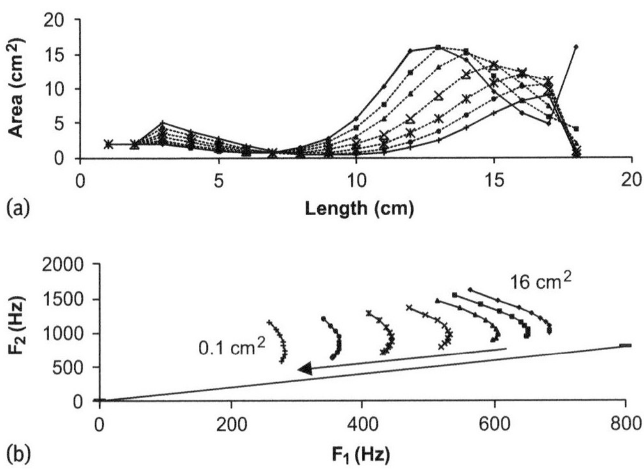

Fig. 3.18: Results obtained with the algorithm for different constant aperture openings (16, 4, 2, 1, 0.5, 0.25, 0.1 cm^2), (a) Area functions after ten iterations for each aperture opening, (b) corresponding trajectories in the F_1-F_2 plane.

Starting from the uniform tube, increasing F_1 and decreasing F_2 lead to the results shown in Figure 3.18 after ten iterations. Acoustic effects are small for apertures between 16 and 2 cm^2. Saturation effects are observed near $F_1 = F_2$.

Figure 3.18 b shows that a rectilinear trajectory is obtained in the F_1-F_2 plane corresponding to a constriction area equal to 0.5 cm^2 moving longitudinally from the back to the center of the tube (Figure 3.18 a, Carré and Mrayati, 1995; Carré, 2009a). Although not shown here, similar results are obtained for a constriction moving longitudinally from the front to the center of the tube.

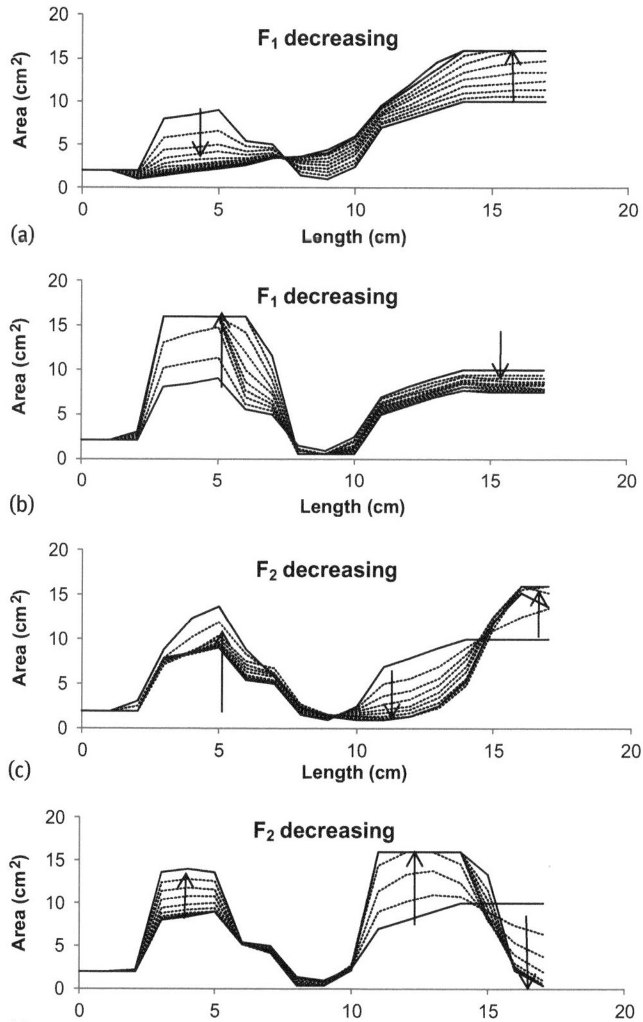

Fig. 3.19: The algorithm is used to increase, decrease F_1 then F_2. Distinctive regions do not appear.

3.2.4 Starting from a non-uniform tube

In all experiments of the present subsection the targeted formant configurations were obtained starting from a uniform tube with an area of 4 cm^2 over its entire length. However, imposing this initial condition is not a prerequisite; the algorithm can be started from any arbitrarily chosen initial state of the tube. In a closed-open tube, starting from any shape obtained during the iterative process (i.e., from an anti-symmetrical shape with the area on a logarithmic scale, see Subsection 3.2.2.1), leads to distinctive regions. Carré, Lindblom, et al., (1995) used a conic shape (anti-symmetrical on a linear scale but quasi-anti-symmetrical on a logarithmic scale) as the initial shape, for which distinctive regions were also obtained.

But, with an arbitrary configuration as initial state (Figure 3.19), the area deformations are complex and they do not predict the distinctive regions obtained when the initial tube state is uniform or anti-symmetrical.

3.3 Conclusions

This chapter showed that efficient deformations along distinctive regions of an acoustic tube 18-cm in length and area functions within the range of 0.5 to 16 cm^2 are obtained from the uniform tube; these deformations can define efficient trajectories and a formant space usable for acoustic communication. Such distinctive regions cannot be obtained from a tube of any arbitrary shape. In the iterative algorithm, a new shape is obtained already at the first iteration by deforming the uniform tube according to the sensitivity function for each formant. Distinctive, unequal regions appear essentially immediately after the first iteration and the subdivision of the tube into these regions remains the same over the entire series of iterations. In fact, the boundaries of the distinctive regions obtained by the algorithm correspond to the zero-crossings of the sensitivity functions $S_i(0, n)$ for a uniform tube. As long as the evolution of the tube is anti-symmetrical, the boundaries of the regions remain stable.

The principles and the characteristics of these deformations (rectilinearity, symmetry, anti-symmetry, synergy, etc.) can constitute the basis of a model that can achieve acoustic production compared to natural speech. Such a model will be presented in detail in the following chapter.

4 The Distinctive Region Model (DRM)

4.1 Introduction

In the previous chapter the use of an iterative computational algorithm (Equations 3.1 and 3.2) based on physical laws and principles led to the conclusion that an 18-cm long tube, when deformed according to a specific strategy, is in fact, an *efficient* and *optimal* tool for acoustic communication. At the cost of *minimum effort* (exerted in minimum cross-sectional deformation of the tube), maximum *acoustic contrast* between two static or dynamic sounds is obtained. *Effort* refers to the energy expanded to *deform* the tube, and *acoustic contrast* represents the resulting changes in the tube's resonance characteristics. A system obtained this way for the *closed-open tube* has important properties, such as: (1) the set of deformations is *efficient* and translates, in the acoustic resonance domain, into linear *formant changes*, (2) efficient F_1-F_2 trajectories are inherently embedded in a dynamic process of resonance modes exploiting the intrinsic acoustic properties of the tube; (3) the area deformations are *transversal*, and obey intrinsic aspects of the tube (anti-symmetry, synergy, linearity, and orthogonality); (4) the tube is deformed at specific spatial *distinctive regions*, in order to arrive at the desired resonance behavior, that is patterns of formant transition behavior; (5) even *large* ranges of area deformations $A(x)$, varying between 0.5 and 16 cm^2, do not affect the boundaries of these regions; (6) the computational algorithm results in *two* distinctive regions for the first formant, *four* for the first and the second combined, and *eight* for the three formants together. Changing the cross-sectional areas in these regions produces specific Formant Transition Behavior (*FTB*) patterns. We will show, consistently with Chapter 3, that the case of closed/quasi-closed tube leads to different regions and *FTB*.

In this chapter we will see that the fixed regions arising as an automatic consequence of the algorithm can be implanted in a model, the Distinctive Region Model *DRM* (Mrayati et al., 1988)[1]. The conclusions of the algorithm of Chapter 3 revealed that boundaries of the regions of the model and their *FTB*s remain valid even for *large* area function changes, as long as the area deformations follow the control strategy mentioned (efficient deformations). When the DRM is controlled by a strategy that obeys the above rules, it is inherently consistent with the cri-

1 In that earlier publication the regions of the model were deduced based on zero crossings of the sensitivity function curves in the neutral tube, and then the distinctive features of the regions were investigated to find cases where they did not change even when the area deformations were *large*.

DOI 10.1515/9781501502019-005

terion of our deductive approach: production of a maximum acoustic contrast through a minimum deformation of the area.

The following sections describe the main aspects of the model: the tube structured in eight *uniform* sections of unequal lengths, *distinctive trajectories* produced by the DRM in the F_1-F_2 or the F_1-F_3 plane, *nomograms* of the model that illustrate the relationship between changes of formant frequencies and changes of area cross-sections in a given region, and the three *distinctive* tube configuration *modes* of the model. As it will be shown, all these aspects are inherent properties or intrinsic consequences of the model's special structure.

4.2 The model

Following the methodology of Chapter 3 that allowed the deduction of fixed distinctive regions as well as efficient $A(n)$ deformation and distinctive trajectories, a model of sound production – the Distinctive Region Model – is formulated. This model consists of simple tube-area structures: N uniform area sections of unequal length. N is 2, or 4, or 8 for the closed-open, and 3 or 7 for the closed/quasi-closed tube. The model incorporates two types of relationships: mapping efficient deformations of $A(n)$ into linear trajectories in the acoustic space, and transforming deformations of distinctive region areas into formant transition behavior (specific *FTB*s). Its command structure is simple, with few transitional controls of region areas. It operates within a maximum acoustic space (see Figures 3.5 and 3.6 in Chapter 3) that permits the exploitation of any sufficient space. It fulfills all principal criteria of the tube: orthogonality, linearity, efficiency, and optimality.

As mentioned before, the model is deduced for two *cases* of tube structure, the closed-open and the closed-closed case. For both cases three *types* of region structure will be presented – types taking into account the first, the first two, or the first three formants. For the closed-open tube the model will yield two, four, or eight distinctive regions, depending on the type, that is whether it includes F_1 only, F_1 and F_2, or F_1, F_2, and F_3. The region deformations can exploit specific acoustic properties, such as anti-symmetry, synergy or compensation, linearity, and pseudo-orthogonality. In the following subsections we will discuss: (1) polarity of the sensitivity functions (*SF*) corresponding to each region for deformations starting from the uniform tube; (2) Formant Transition Behavior (*FTB*) of the regions; (3) nomograms for each formant i as a function of the cross-section area of tube region n, that is $F_i = f[A(n)]$; and (4) the linearity of the dynamic acoustic trajectories in the F_1-F_2 plane corresponding to distinctive efficient deformations. The three structure types of the DRM will be introduced for the closed open and closed-closed cases, although they will be presented in detail for the first case only.

4.2.1 The two-region model for the closed open case

Based on the tube's properties described in Chapter 3, the model structure in which only the first formant F_1 is controlled (although $F_2, F_3, \ldots$ also do exist) identifies two regions. As illustrated in Figure 3.8 of the previous chapter, the iterative process yields two distinct regions and large changes (proportional to the sensitivity function) are observed in the cross-sectional areas that produce large, efficient, and linear increments or decrements in F_1. An increase in the area at any location in the front half of the tube raises the frequency of the first formant, and an area increase in the back half lowers it. This two-region type leads to a model of two rectangular regions labeled R_1 and R_2 with a boundary at the midpoint of

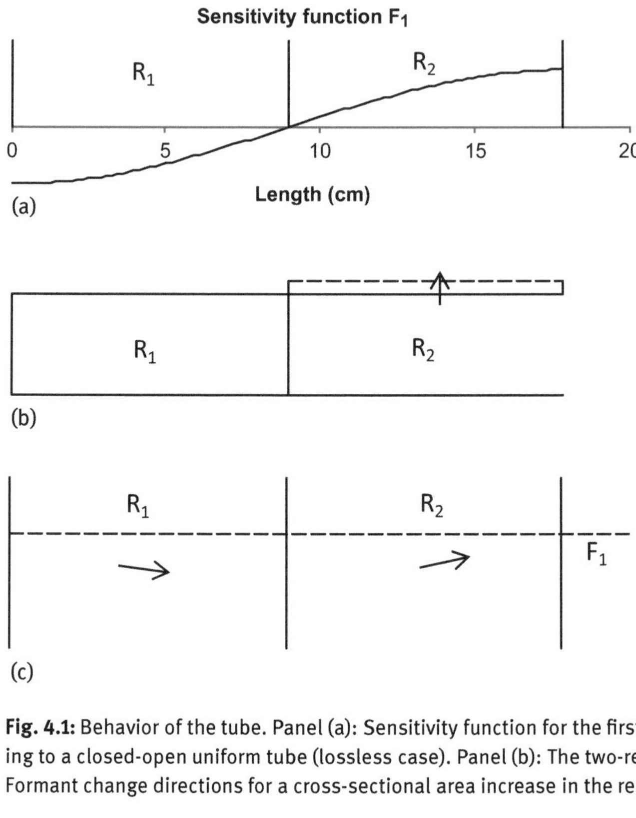

Fig. 4.1: Behavior of the tube. Panel (a): Sensitivity function for the first formant corresponding to a closed-open uniform tube (lossless case). Panel (b): The two-region model. Panel (c): Formant change directions for a cross-sectional area increase in the regions of Panel (b).

the tube (corresponding to the zero-crossing of the first formant's sensitivity function) and exhibits specific *FTB*: specific changes in the uniform area of the two regions lead to efficient and monotonic first formant changes. The area change is obtained by use of a transversal command along the effective length L of the tube (see Chapter 3, Subsection 3.2.2). As Figure 4.1 shows, the front and back regions

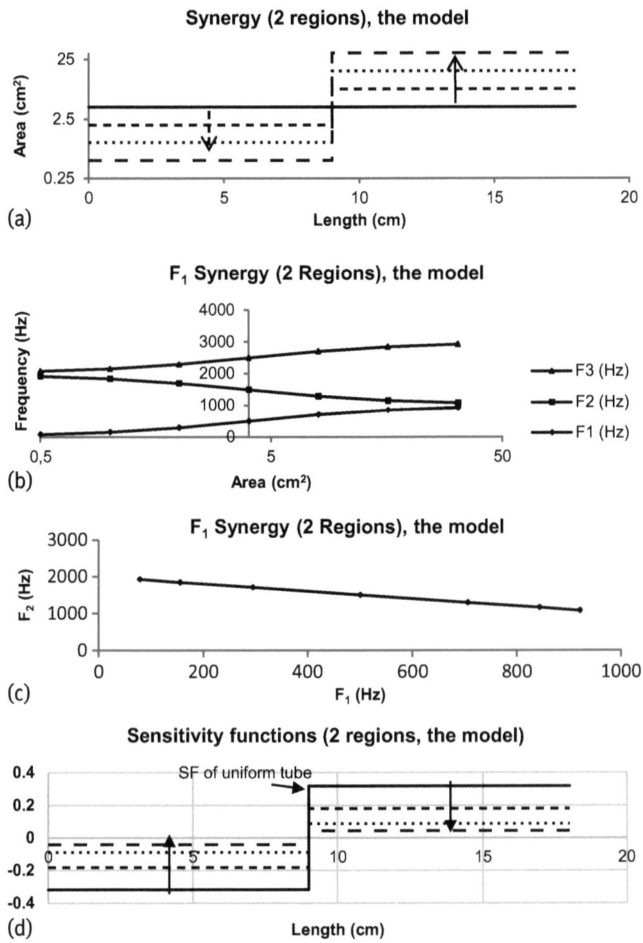

Fig. 4.2: Behavior of the model: Formant frequency variations for the three formants with the cross-sectional area varied from 0.5 to 32 cm^2 when the tube is configured as 2 sections of 9 cm each (the model). Panel (a): "efficient" deformation of regions when increasing F_1 starting from the neutral configuration; Panel (b): corresponding changes in the three formants (increasing and decreasing F_1); Panel (c): resulting efficient linear trajectory in the F_1-F_2 plane. Panel (d) the corresponding sensitivity functions when increasing F_1.

exhibit an anti-symmetrical behavior – an increase of the deformation in the front region will acoustically compensate for a similar increase in the back region and the area changes of the two regions in opposite directions reinforce each other and thus result in a maximal formant change (i.e., they exploit synergy).

Figure 4.2 shows behavior of the model relating formant frequency variations to that of cross-sectional areas, and the corresponding sensitivity function. Synergy is applied, that is the section-by-section product of areas under the two regions in the model (they have the same value all along the region), $S(R_1)$ and $S(R_2)$, is held constant (16 here). The area varied from 0.5 to 32 cm^2 in logarithmic steps. Owing to the synergetic area modification, the amount of the formant change is multiplied by 2. If we use this efficient area deformation as a command for the model, then we actually dispose of a single command because the change commands (up or down) in the two regions are not independent, meaning the number of control parameters in the model is divided by two (see Figure 3.8 in Chapter 3). Efficiency ($\Delta F_1/\Delta A$) is *maximum* around the *neutral configuration.*

It is important to note the effect of the model structuring the 18 1-cm tube sections into a two-region tube (and into four- and eight-region tubes, as shown later). Figure 4.3 illustrates the synergistic changes in the area function and the corresponding acoustic results, when the tube is quantized into 18 equal sections rather than only two regions. Comparing Figure 4.2 (i.e., the model) with Figure 4.3 (i.e., the theoretical case) reveals certain differences: firstly, when we control F_1, formants F_2 and F_3 remain stable in Figure 4.3 but exhibit a minor change in Figure 4.2, secondly, compared to Figure 4.3, the control in Figure 4.2 is very simple, thirdly, the sensitivity function is a step-function in Figure 4.2 but sinusoidal in Figure 4.3. Nevertheless, the range and the behavior of F_1 remain similar in the two cases.

4.2.2 The four-region model for the closed-open case

When both F_1 and F_2 are controlled (F_3 exists but it is not controlled), the model acquires a four-region structure. Recalling Figure 3.10 in the previous chapter, the iterative process there yielded four distinct regions and we saw large changes in the cross-sectional areas that produced a large, efficient, and linear increase or decrease in F_2. In contrast to the two-region case, both the front and the back regions are divided into two unequal ones, and thus four regions are obtained with boundaries coinciding with those of zero-crossings of the sensitivity function of F_2.

Combining the previous F_1-bound and this F_2-bound region leads to creation of the four-region model with regions labeled R_1, R_2, R_3, and R_4. Figure 4.4 sum-

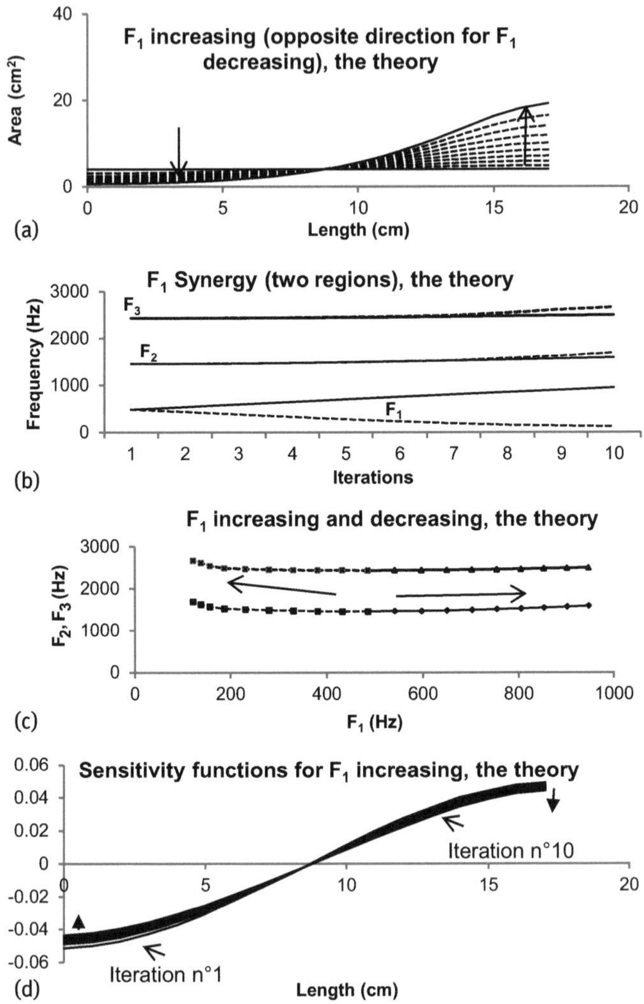

Fig. 4.3: Case similar to Figure 4.2, reproduced from Chapter 3, Figure 3.8: frequency variations of the three formants with the cross-sectional area varied from 0.5 to 16 cm^2 when the tube is configured as 18 sections of 1 cm length each. Panel (a): efficient deformation of regions for F_1 increasing, but for F_1 decreasing it is in the opposite direction; panel (b): corresponding changes in the three formants for increasing and decreasing F_1; panel (c): resulting efficient linear trajectory in the F_1-F_2 plane; panel (d) corresponding behavioral variation of the sensitivity function changes when increasing F_1.

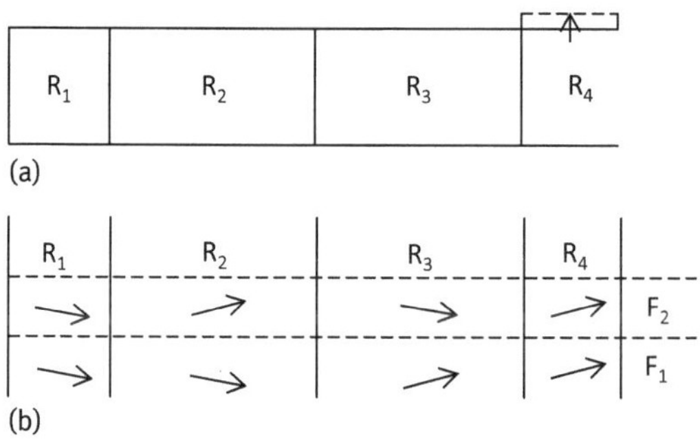

Fig. 4.4: Panel (a): the four-region model. Panel (b): The formant variation directions when the section areas in each region increases.

marizes the FTB in each region. With L being the effective length of the tube, the lengths of the regions are $L/6, L/3, L/3, L/6$. The anti-symmetry property observed in the two-region model is preserved here, too, and pseudo-orthogonality of the direction of formant changes can also be seen: each of the four regions carries one of the four combinations of the rising or falling direction of the F_1 and F_2 changes.

Figure 4.5 illustrates nomograms and the corresponding average sensitivity function value changes obtained when the cross-sectional area of each of the four regions (R_1, R_2, R_3, and R_4) varies from 0.05 to 16 cm^2 *in logarithmic steps*. As seen in panels (a) and (c), the ranges of both F_1 and F_2 changes are most efficient inside the 0.5 to 16 cm^2 range, consistent with what has already been shown for the two-region system. This range is obtained starting from the uniform tube the cross-sectional area of which, as stated earlier, is always 4 cm^2. Panels (b) and (d) show that the maximum sensitivity for F_2 is at an area of about 4 cm^2, that is around the neutral. For F_1 the maximum sensitivity value changes across the regions. When the average value of the sensitivity function of a region is zero the frequency nomogram curve is maximum.

4.2.3 The eight-region-model for the closed open case

In a way similar to the above description of the two- and the four-region types, in the following subsections regions, nomograms, and modes of the more frequently used eight-region structure will be presented.

4.2.3.1 Regions of the eight-region model

Chapter 3 already demonstrated that, when taking F_3 into consideration, six regions can be seen (see Figure 3.12). Combining the first three formants, a model with eight regions emerges. Figure 4.6 a shows these regions, labeled

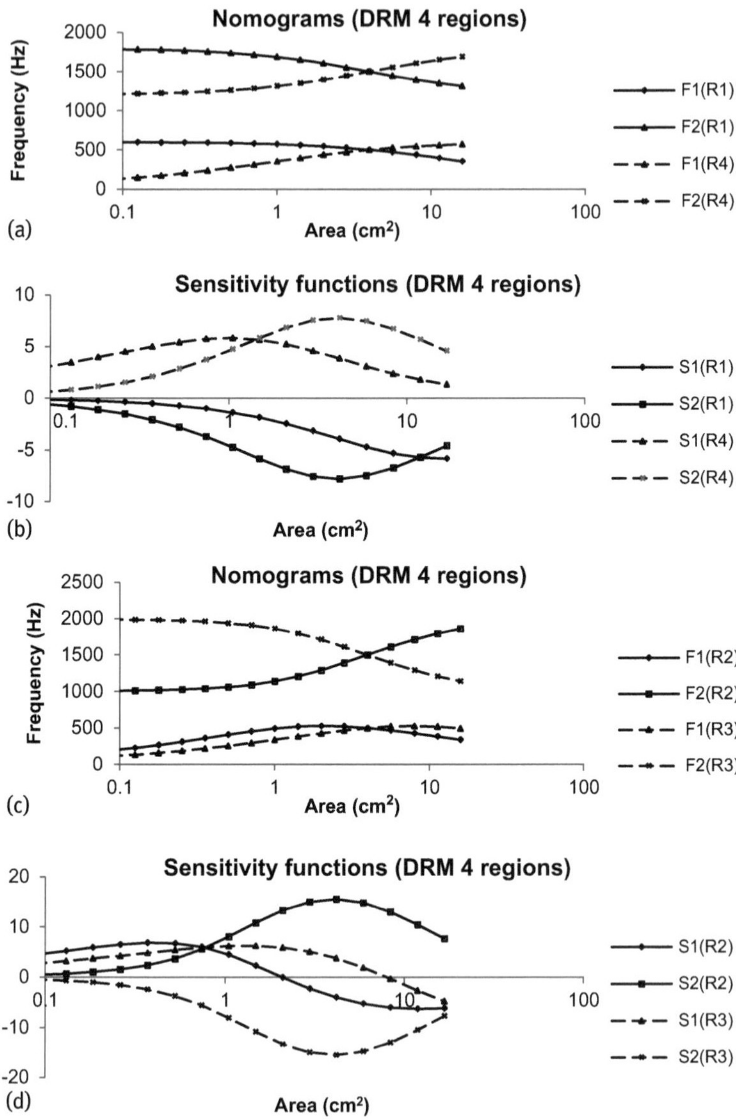

Fig. 4.5: Panels (a) and (c): Nomograms of the four-region DRM model for the closed open case, showing changes in F_1 and F_2 when region areas are changed between 0.05 and 16 cm^2. Panels (b) and (d): Corresponding change in average sensitivity function values of each region.

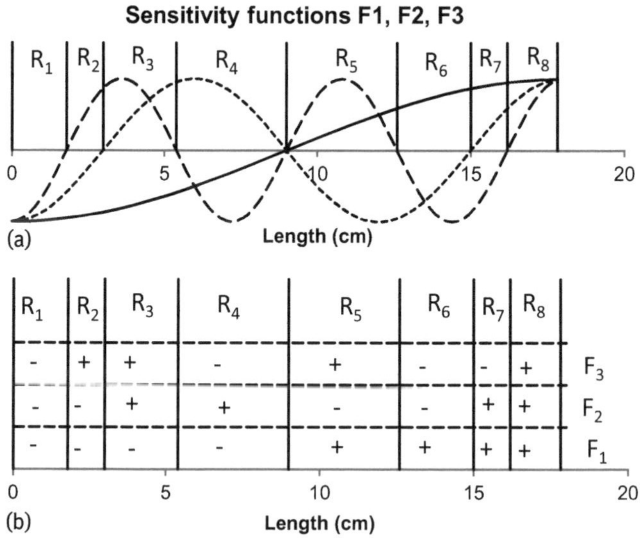

Fig. 4.6: Sensitivity functions for the three formants in the case of a uniform tube and the eight regions; (b) Corresponding polarity matrix (for example for R_8, the three formant sensitivity functions are positive). The eight regions have eight (= 2^3) distinct *FTB*s.

$R_1, R_2, R_3, R_4, R_5, R_6, R_7$, and R_8 going from the closed to the open end of the uniform tube. Region lengths coincide with the zero-crossings of the superimposed sensitivity functions of the three formants, that is for an effective tube length L the lengths of the eight regions will be $L/10, L/15, 2L/15, L/5, L/5, 2L/15, L/15$, and $L/10$. Figure 4.6 b shows the polarity matrix of the sensitivity functions. The eight regions have 2^3 distinct combinations of polarities, i.e., Formant Transition Behaviors (*FTB*s).

Figure 4.7 displays rising and falling *FTB* patterns for the eight regions. For instance, in the R_8 region, a cross-sectional area increase results in the rise of all three formant frequencies. Similarly, the R_7 region displays an *FTB* in which an increase in the area will cause a rise in F_1 and F_2 but a fall in F_3. Following this scheme, a specific *FTB* can be clearly defined for each of the eight tube regions.

*FTB*s of the eight regions thus defined are valid in tube configurations in which a *coupling* between back and front cavities exists for cross-sectional areas larger than 0.5 cm^2. The variation of the areas used, between 0.5 and 16 cm^2, spans a five-octave range. Subsequently, this condition will be referred to as the One-Tract Mode (see Subsection 4.2.3.2 below).

As deduced in Chapter 3 from tube acoustics for *each* formant, as long as the tube is anti-symmetrical about the midpoint and efficient deformations are ap-

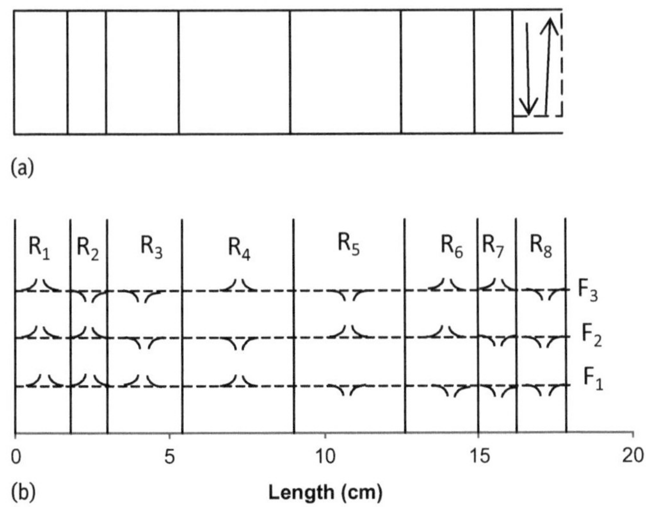

Fig. 4.7: Formant transitions obtained by successively closing and opening each of the eight regions from and to a uniform tube shape. Panel (a): The eight regions with the contracting and expanding area change illustrated in R_8 for reference. Panel (b): diagrams of formant frequency change patterns (Formant Transition Behaviors) in each region resulting from an area reduction followed by an area expansion. The eight regions have 2^3 distinct *FTB*s.

plied to it, region boundaries will coincide with the zero-crossings of the sensitivity function *SF* that are inherently stable and, therefore, the boundaries will be necessarily *constant*. As a consequence, in tube zones near region boundaries formants will not be sensitive to area changes.

For tube shapes that are *not anti-symmetrical* and that contain no points of tight constriction, *SF* zero crossings will exhibit a *slight* local shift within a given zone. An example of this phenomenon can be seen in Section 3.2.3 of Chapter 3 showing what happens when the source cavity is held constant at the input of the tube. Figure 3.16 displays the slight local deviation of this zero-crossing for F_1, and Figure 3.17 for F_2). As clearly indicated in Figure 4.6, there is the possibility for one local zero-crossing deviation of F_1, three of F_2, and five of F_3.

Finally, in situations where the fourth formant would also be controlled, the need to define 14 regions each with its proper *FTB* is easy to demonstrate. In general, the number of regions is defined by the relation Number-of-Regions $= 2 \times (1 + (1 + 2 + \ldots + m - 1))$ where m is the number of formants.

Formant transition behavior FTB for each region of the eight-region model is distinctive. Figure 4.7 illustrates *FTB* changes for each region when the region's cross-sectional area is first decreased and then increased. Clearly, the behavior of regions is anti-symmetrical about the mid-point of the tube: each region in the front half of the tube has its reverse counterpart in the back half of the tube. *FTB*s of the eight regions of the uniform tube illustrated in Figure 4.6 show that all 2^3 possible combinations of the three formants are present. This means that, when three formants are taken into account, this system incorporates *all* possible codes using rising or falling formant transitions. Thus, also from this point of view, the system is *optimal.*

Looking at Figures 3.8, 3.10, and 3.12 (Chapter 3) from the model's standpoint, area changes in certain regions or region combinations result in *linear* formant variations. Such linear behavior applies to a formant whenever an area change takes place in a region or in a combination of regions between two zero-crossings of the sensitivity function for that formant, or by one zero-crossing following the closed or preceding the open end. Regions that exhibit linear behavior can be combined and the resulting effect is also linear. Thus, in the eight-region model the following region combinations exist: for F_1, the regions $R_1 + R_2 + R_3 + R_4$ and/or their front counterparts $R_5 + R_6 + R_7 + R_8$; for F_2, the regions $R_1 + R_2$ and/or $R_3 + R_4$ and/or their front counterparts $R_8 + R_7$ and/or $R_6 + R_5$; and for F_3, the regions R_1 and/or $R_2 + R_3$ and/or R_4 and/or their front counterparts R_8 and/or $R_7 + R_6$ and/or R_5. The and/or means that, if desired, we can have linear synergy or linear compensation.

4.2.3.2 Distinctive modes of the eight-region model

Tube configurations can be classified into three distinct categories depending on the degree of narrowing of the cross-section area of regions, namely, finite non-tight constriction, complete closure, or a transitional category between the two. Each category of configurations distinguishes a mode of operation of the tube. They are, respectively, the One Tract Mode (*OTM*), the Transition Mode (*TM*), and the Two Tract Mode (*TTM*) (Figure 4.8 a). The *OTM* is valid for constrictions down to about 0.5 cm^2, the *TM* is valid between about 0.5 and about 0.01 cm^2, while the *TTM* is valid for constrictions narrower than 0.01 cm^2.

- In the *One Tract Mode* (*OTM*), the tube is considered as a single unit with all regions tightly coupled. The boundaries of the regions are fixed as long as we apply an *efficient deformation* around the neutral, that is, synergy and anti-symmetry (see Figures 4.2 and 4.3 and Figures 3.8, 3.10 and 3.14 in Chapter 3). They move slightly, or "jitter," when these two conditions are not met, but the jitter zones are narrow (see Chapter 2, Figure 2.7, and Figures 4.16 and 4.17 where we have a fixed source cavity and a non-neutral tube respectively. As Figures 4.8 and 4.9 illustrate, in this mode most regions conserve their characteristics, that is, their *FTB*s follow their behavior as displayed in Figure 4.7, although some regions do not consistently do so over the whole *OTM* range. As demonstrated in Chapter 3, it is mainly in the *OTM* mode that efficient deformation prevails and, consequently, efficient formant transitions take place.

- In the *Two Tract Mode* (*TTM*), the tube momentarily breaks into two *uncoupled* parts as the result of a constriction closing a region. As already mentioned in Subsection 4.2.3.1, regions of the model where constrictions are introduced are the ones that produce the optimal number of distinctive *FTB*s – 2^3 for three formants. This mode is essential for the generation of dynamic sounds. Examples showing ways in which this is accomplished will be presented and explained in Chapter 7. When the constriction occurs in region R_8, we have only one back and no front tube, and thus the originally closed-open tube becomes closed-closed – a case to be discussed in a subsequent section of the present chapter.
- When shifting from a *TTM* to an *OTM* configuration or vice versa, one needs to traverse the *Transition Mode* (*TM*). In the *TM*, zero-crossings shift their location and, consequently, the regions change as a function of the degree of constriction. This is clear because the *distinctive regions* are defined strictly only for the *OTM*.

4.2.3.3 Nomograms of the eight-region model

Figure 4.8 displays nomograms $F_i = f[A(n)]$ for a closed-open tube. They numerically express the relation between the cross-sectional area under each of the eight regions and the three formant frequencies. The range where each of the three modes *OTM*, *TM*, and *TTM* prevails is indicated. When the area of any region changes, it is varied between 0.01 and 16 cm^2 on a logarithmic scale, with areas of all other regions held constant at 4 cm^2. The nomograms have been calculated including the effects of loss, particularly the one due to wall vibration.

Looking at the four nomogram panels in Figures 4.8 and 4.9, we would like to point out that, in the *OTM* illustrated, the extent of formant frequency change caused by area changes differs across regions. The formant change is particularly small in region R_7 (see the second panel). Also, the area span of the three modes is not identical in all regions and the behavior of certain formants in some regions changes very little at the boundaries of the mode (see F_1 for R_4, in the fourth panel). Note also that the range of variation of formants F_1 and F_2 for regions R_2 and R_7 do not affect the *OTM*, and consequently they cannot be efficiently controlled.

The model displays *pseudo-orthogonality features* of the eight regions and *compensatory/synergetic* aspects between them. This orthogonality is particularly important in the four-region model, because those four regions determine the F_1 and F_2 behavior in the F_1-F_2 plane. This fact sheds light on the triangular shape of vocalic space in the F_1-F_2 plane (see Figure 4.10). On the one hand, changes in

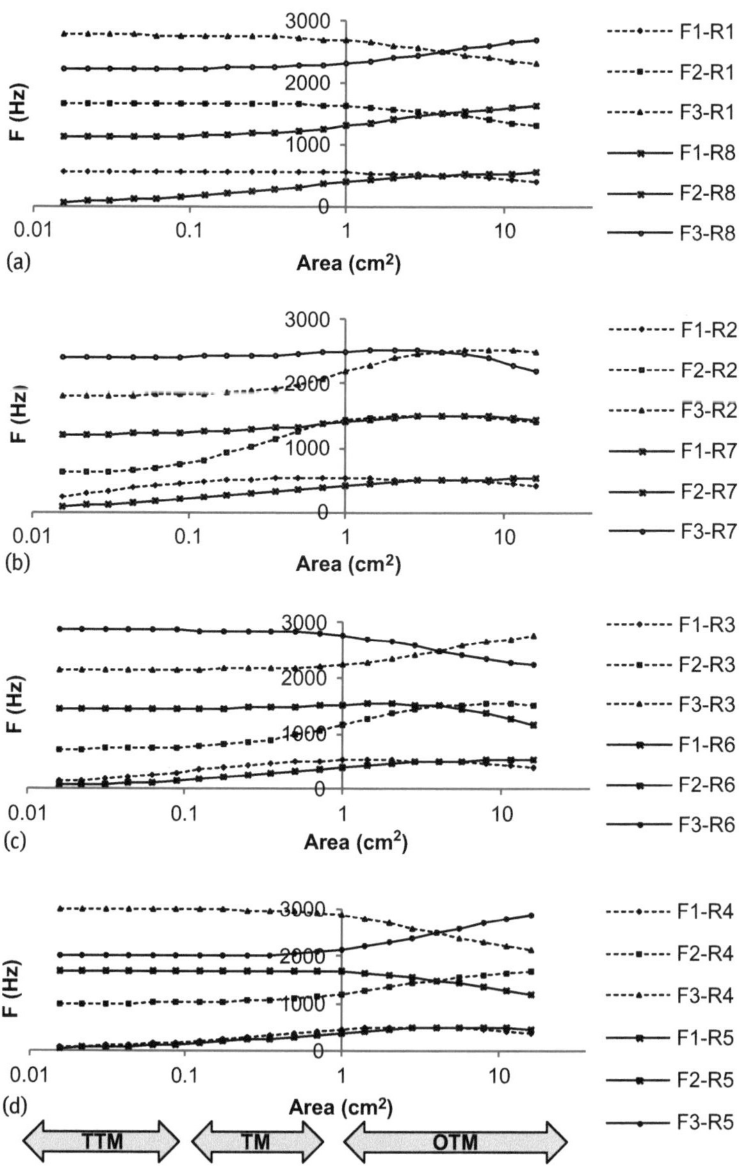

Fig. 4.8: Nomograms obtained for a uniform tube when varying the cross-sectional area of each of the eight regions, one at a time. The area is varied logarithmically from 0.01 to 16 cm^2. The *OTM* prevails up to a constriction of about 0.5 cm^2, while the *TM* prevails from around 0.5 down to 0.1 cm^2 where the *TTM* starts. Each Panel shows two anti-symmetric regions. The top Panel corresponds to R_1 and R_8; the second from top R_2 and R_7; the third from top R_3 and R_6; and the bottom R_4 and R_5.

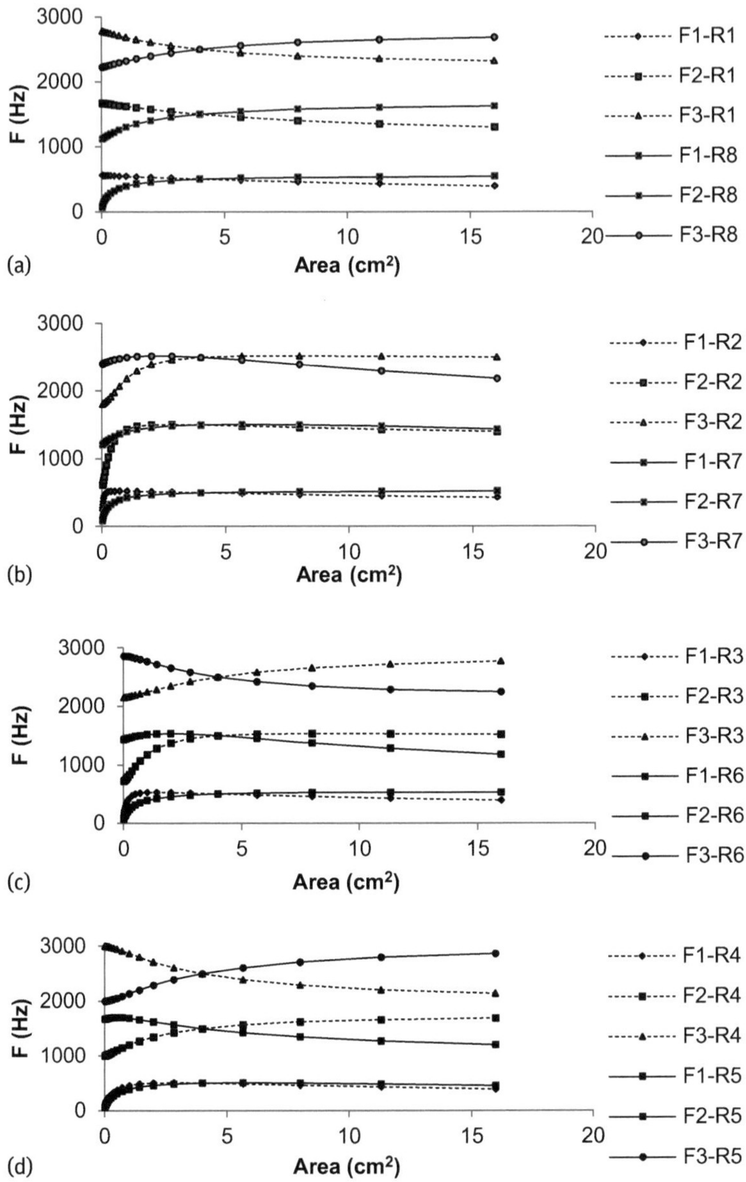

Fig. 4.9: The same nomograms as in Figure 4.8 a plotted on a linear scale. Using this scale, it can be clearly seen that small absolute changes for small cross-sectional areas have the same acoustic effect as large area changes when the area is large (changes are represented as the distance between neighboring points on the curves).

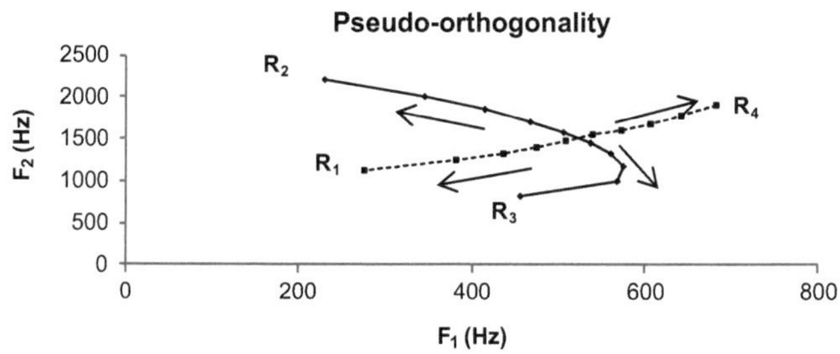

Fig. 4.10: Pseudo-orthogonality shown for the case of the four-region model presented in the F_1-F_2 plane. The increase of the cross-sectional area of each of the regions produces a change that is pseudo-orthogonal to the effect of the other area change.

R_1 or R_4 lead to an acoustically expected covariation of F_1 and F_2 while, on the other hand, changes in R_2 or R_3 lead to opposing F_1-F_2 variations, that is, when one increases the other decreases. Similar pseudo-orthogonal aspects are present in the F_1-F_3 and F_2-F_3 planes.

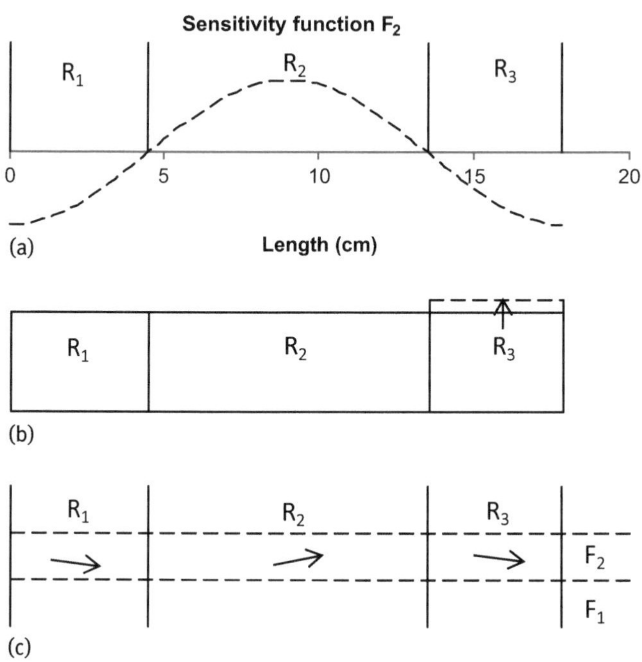

Fig. 4.11: Spatial regions of the *first two formants* of a uniform tube closed at both ends (lossless case), the corresponding region model, and the matrix of the formant change directions for region area increases.

4.2.4 The closed-closed tube model

The tube closed at both ends was also deduced using the iterative algorithm in Chapter 3. As explained there, the number of regions in the closed-closed case is equal to $(2^n - 1)$ where n is the number of formants considered. Just as for the closed-open case, region boundaries continue to coincide with zero crossings of the sensitivity functions of the uniform tube. In the closed-closed tube,

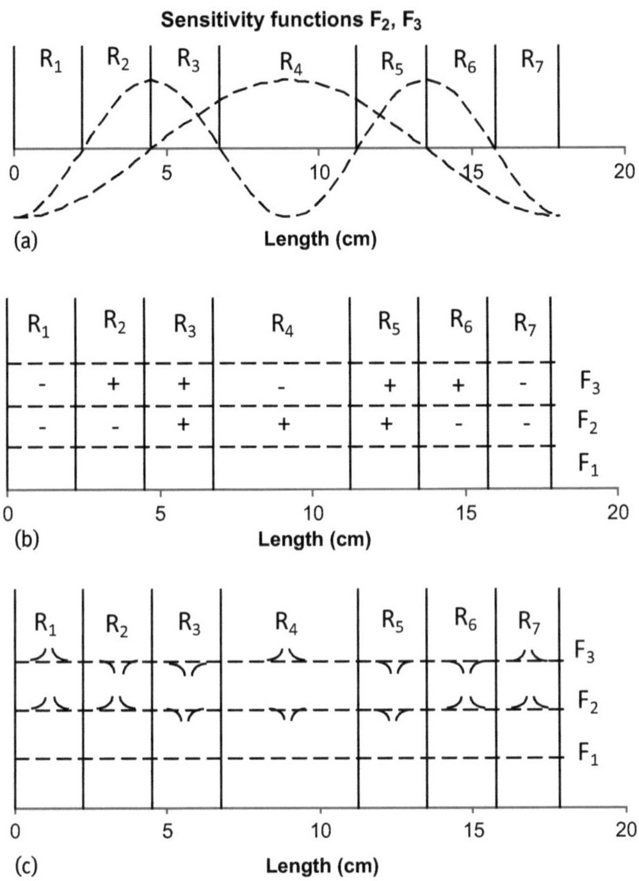

Fig. 4.12: Panel (a): Spatial regions obtained from the sensitivity functions considering *three formants* in a uniform tube closed at both ends (lossless case) and the corresponding sensitivity functions. Panel (b): The matrix of formant change polarities for increases of the area of each region. Panel (c): diagrams of formant frequency change patterns in each region resulting from an area reduction followed by an area expansion. There are four different *FTB* for seven regions.

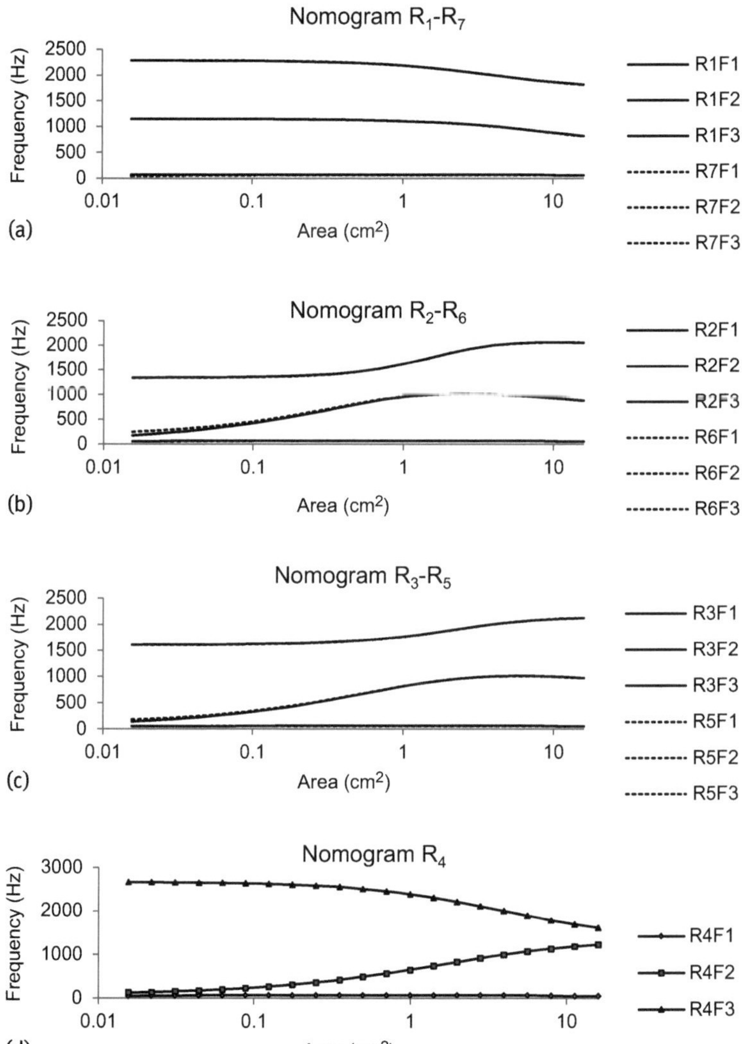

Fig. 4.13: Nomograms of the seven-region closed-closed tube when varying the area of the regions between 0.05 and 16 cm^2 on a logarithmic scale. The three top panels show the nomograms of symmetric regions. The nomogram in the bottom panel is that of the central region (R_4) alone; it is shown because it has no symmetric pair. In each of the top three panels, regions were chosen in order to illustrate that nomograms of symmetric regions (R_1-R_7, R_2-R_6, R_3-R_5) are identical.

the pseudo-orthogonal behavior is not preserved and, instead of anti-symmetry, a symmetrical behavior is observed. This section will briefly describe the model's treatment of the closed-closed tube both for the three- and the seven-region configurations that correspond to taking into account two or three formants, respectively. Figure 4.11 shows the three regions for the *two-formant* case. For F_l the sensitivity function is zero and for a tube with rigid walls the frequency of the first formant is also zero. (For a tube with yielding walls the value corresponds to the wall's mechanical resonance frequency.) Here, the behavior of the model is symmetrical and the lowest F_2 is obtained when a synergetic command is applied to R_2 (decreasing), and to both R_l and R_3 (increasing).

When *three formants* are taken into account, seven regions are defined. Figure 4.12 shows the same seven regions as deduced in Chapter 3. The figure illustrates for F_2 and F_3 the regions bounded by zero-crossings of the sensitivity functions as well as the F_2 and F_3 sensitivity functions including their polarities. Figure 4.13 shows the corresponding nomograms.

4.3 Use of the model to discover primitive trajectories

4.3.1 The model with simplified controls

The closed-open tube model with *eight regions* offers an ideal environment for precise control of deformations over the entire length of the tube because, as demonstrated in Section 4.2.3, the eight regions represent optimal tube segments. The eight-region model permits the command of the three formants F_1, F_2, and F_3. It also allows the formation of a fixed cavity at the source of excitation, and it locates the places of articulation of consonants, as explained in Chapter 3, Section 3.2.3. In addition, the granularity of deformations is significantly lower than that of a four-region model and, consequently, it approximates a more realistic sound production system. On the other hand, control of the eight-region model is more complex than that of the four-region model because the number of deformation commands in the former is twice the number of commands in the latter (see Chapter 3). Thus, in the following subsection we will introduce specific simplifications to the eight-region model along the following scheme: (1) Area of region R_1 will be fixed and equal to $2.5\,\text{cm}^2$ (approximating the size of the source, the larynx cavity), the cross-sectional area of R_8 can become sufficiently narrow when the closed-open tube is made to approach a closed/quasi-closed one, while it can become sufficiently wide when a closed/quasi-closed tube is made to approach a closed-open one; (2) the closed/quasi-closed configuration with a central constriction is approximated by narrowing regions R_4 and R_5; (3) the total volume of regions R_3, R_4, R_5, and R_6 is kept *practically* constant, in order to avoid the cross-sectional area becoming unrealistically large; (4) the cross-sectional area of R_2 will be the average of those of R_1 and R_3, and the area of R_7 the average of those of R_6 and R_8; (5) the cross-sectional areas are set to vary within the range of 0.5 to 10 (sometimes 16 cm^2) thus encompassing a range of over four octaves; (6) the commands used can be either transversal or longitudinal.

The resulting simplified model has three types of commands: the first to control the *place and degree of the constriction*, the second to control the *aperture*, and the third to control *closing a given region*, as illustrated in the nomograms (Figure 4.8). The first two commands produce the seven distinctive trajectories in the F_1-F_2 plane explained in the following sections, while the third produces eight F_1-F_2-F_3 formant transitions based on *FTBs* (Figure 4.7). Throughout the book, in all figures the constriction displacement command is represented by solid and the aperture command by dotted arrows.

Figures 4.14 and 4.15 show the simplified model with its transversal commands directed at the place and degree of constriction and the aperture. This model will be used in all following chapters.

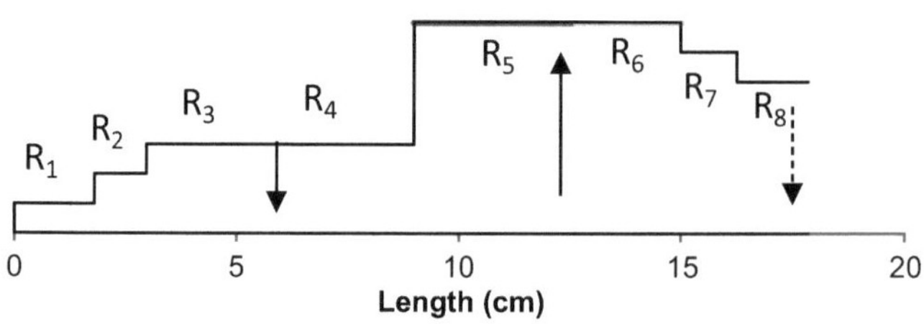

Fig. 4.14: The simplified eight-region model, closed-open case. The R_1 area is fixed at 2.5 cm^2, R_2 and R_7 areas are automatically assigned the average of the two adjacent regions; only two commands are involved: a change in the R_8 area represents the aperture command, and the synergic $(R_3 + R_4)/(R_5 + R_6)$ command shifts the constriction front to back and back to front.

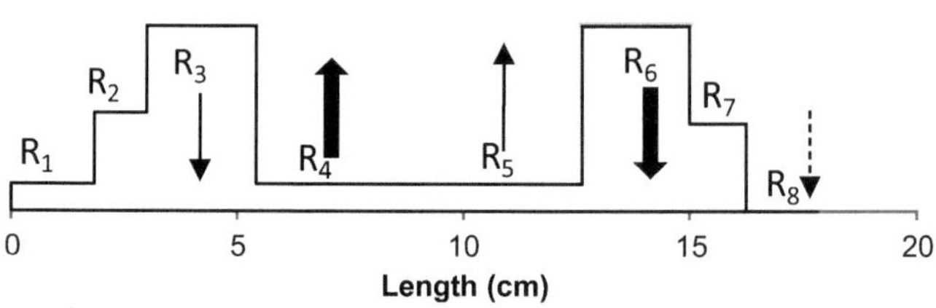

Fig. 4.15: The simplified eight-region model, closed/quasi-closed case. The R_1 area is fixed at 2.5 cm^2, the R_2 and R_7 areas are the average of the two adjacent regions, the R_8 area is quasi-closed and can be controlled; two transversal commands are involved, they are the quasi-synergic commands R_3/R_5 and R_4/R_6 that shift the constriction from central to back (thin arrow) and central to front (thick arrow).

4.3.2 From a closed-open to a closed-closed model

The use of the closed-open and closed-closed models increases the number of possibilities both for the production of sounds and for acoustic trajectories in the F_1-F_2 plane. Producing sounds with low F_1 requires a narrow output aperture and the closed-closed model does offer that. Nevertheless, the closed-open model can also be used with a narrow output aperture (R_8 quasi-closed), and the closed-closed model with the R_8 aperture open.

Building on the results obtained in Chapter 3 (Section 3.2.3.3), the move from closed-open to closed-closed may be achieved by longitudinally displacing the constriction. Figure 4.16 shows the two possible manners for achieving displacement of a front constriction to a central one: either by one longitudinal gesture (L),

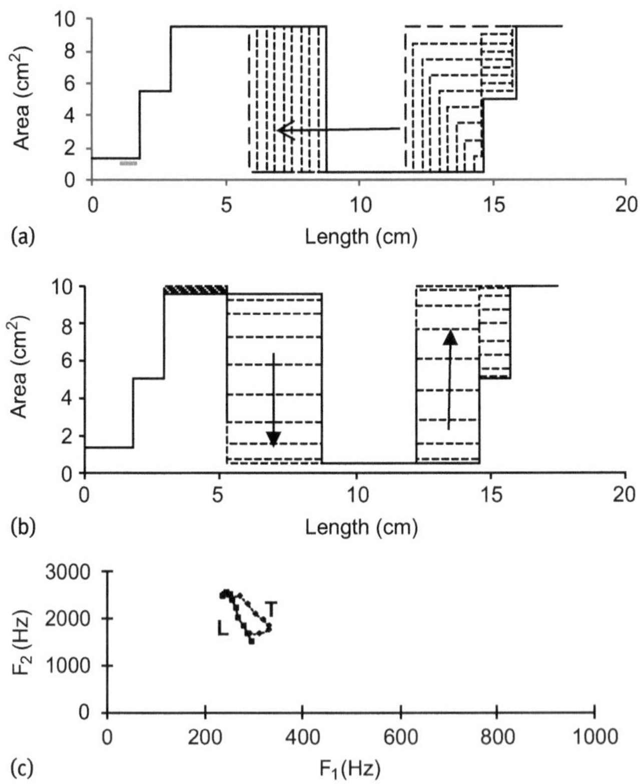

Fig. 4.16: Panel (a): Longitudinal displacement of the constriction from front to central; panel (b): transversal displacement of the constriction; panel (c): corresponding formant trajectories for longitudinal (L) and transversal (T) displacement of the constriction.

or by two synchronous transversal gestures (T). Longitudinal displacement of the constriction yields a more linear formant trajectory than the transversal one.

4.3.3 Deduction of seven primitive trajectories

Using the *simplified DRM* model, with its three commands, seven distinctive trajectories are obtained. These trajectories are the result of applying the first two commands on the closed-open model with the R_8 aperture open or quasi-closed and the closed-closed model with the R_8 aperture either quasi-closed or open. Three trajectories are produced by one of these two commands alone, whereas the four others need both commands for their production. One of the major postulates of our model is that these trajectories constitute primitives of a communication system.

When moving from one point to another, there is an infinite number of area deformations involving an infinite number of formant trajectories across the F_1/F_2 plane (Carré et al., 2001). The seven specific trajectories we are focusing on are products of the simplified model, they carry information on the dynamic acoustic behavior of the tube. They structure the acoustic F_1/F_2 plane, just as the area deformations of the distinctive regions structure the tube shapes. The seven trajectories are illustrated in Figure 4.20.

Figure 4.17(a) shows deformation commands (one or two) applied to the simplified model. The deformation from a configuration with back constriction and front cavity is transversal: it produces a front constriction and a back cavity (antisymmetrical behavior). The corresponding formant trajectory marked by the two ends [4]-[2] in Figure 4.17(b) is rectilinear, except for the curvature in the first three points (from [4] to [1]) that can be explained by non-linear effects. The trajectory [1]-[2] (an increase of F_2 and a decrease of F_1) is linear: it stops at the point beyond which tube deformation is no longer acoustically efficient. When the corresponding constriction command is associated with the R_8 command (aperture quasi-closing), the trajectory [1]-[3] is obtained.

To achieve a rectilinear trajectory (Carré and Mrayati, 1991), we must either anticipate the "second command" in the time domain or apply a logarithmic scale on which the command is produced. The trajectory [1]-[3] then becomes rectilinear and the acoustic effects for each command are equal during their production in the time domain (Carré and Divenyi, 2000).

Figure 4.18 illustrates the way the trajectory from point [4] to [5] is obtained when a closed-open model transitions into a closed-closed model by way of a *longitudinal* "first command" (a longitudinal displacement of the constriction that has an area of $0.5\ \text{cm}^2$) from the back to the center of the tube, co-produced

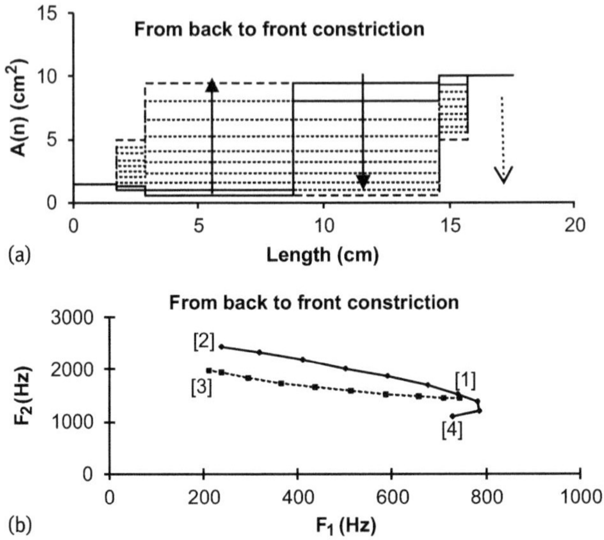

Fig. 4.17: Panel (a): The arrows correspond to the two commands of the simplified closed-open *DRM* model: the bold arrow represents the "first command" using synergy (controlling $R_3 + R_4$ and $R_5 + R_6$ synergically), and the dotted arrows represent the "second command" controlling R_8. Panel (b): Corresponding formant trajectories in the F_1-F_2 plane: the trajectory [1]-[2] is obtained with the "first command," the trajectory [1]-[3] is obtained with "two commands" co-produced.

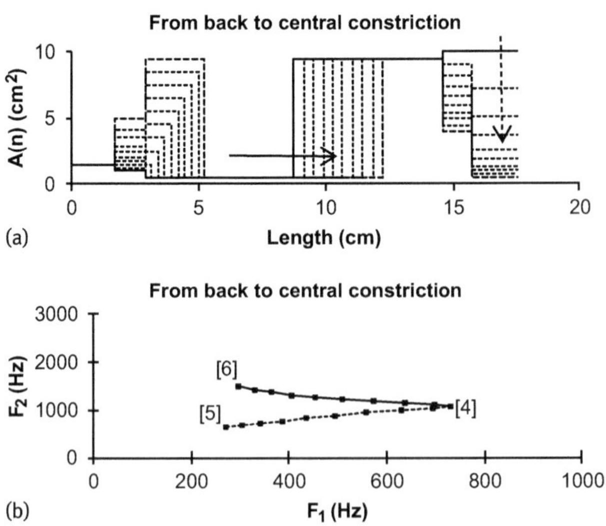

Fig. 4.18: Panel (a): Longitudinal displacement of the constriction from back to central; panel (b): corresponding formant trajectories ([4]-[6] without "aperture" gesture, [4]-[5] with "aperture" gesture).

with a "second command" (i.e., aperture closing). The trajectory [4]-[6] is obtained without "aperture" closure. We want to make it clear that the symbols [4] in Figures 4.17 b and 4.18 b refer to the same point.

The same approach may be used to study the trajectory [2]-[5] (Figure 4.19). The constriction moves longitudinally from the front to the center of the tube and the "first command" is co-produced with the "second command." For the trajectory [2]-[6], the open-end aperture is 10 cm^2. Similarly, for producing the [3]-[5] trajectory the aperture is 0.5 cm^2 and the constriction moves longitudinally from the front to the center of the tube.

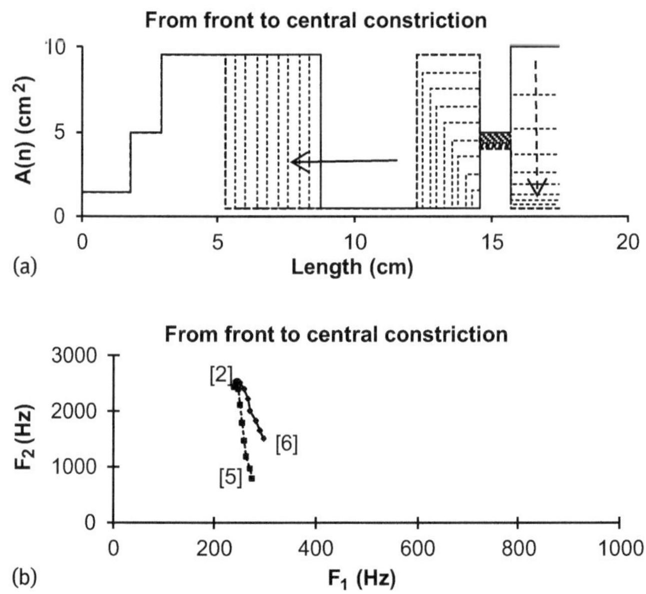

Fig. 4.19: Panel (a): Longitudinal displacement of the constriction from front to central; panel (b): corresponding formant trajectories ([2]-[6] without "aperture" gesture, [2]-[5] with "aperture" gesture).

In summary, the first command (the constriction displacement) leads to three main acoustic trajectories (solid lines in Figure 4.20). Combining the first and second commands leads to four other complementary trajectories (dotted lines in Figure 4.20). Thus, Figure 4.20 shows the efficient capacity of the model in terms of dynamic trajectories rather than in terms of static positions. Each formant trajectory is obtained by means of only a single, or by two co-produced commands representing the simplified control of the DRM.

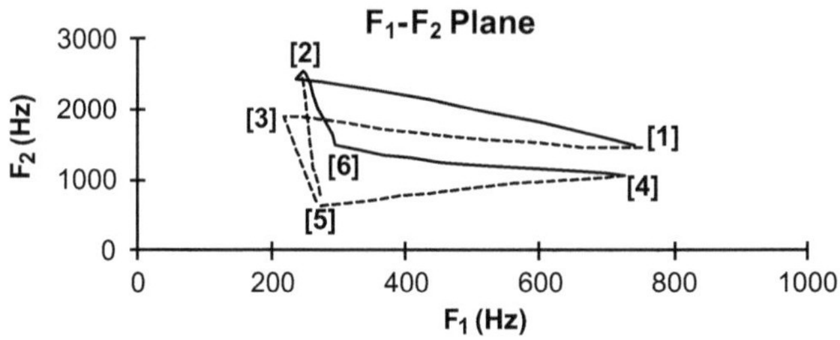

Fig. 4.20: The distinctive primitive formant trajectories obtained with the DRM.

4.4 Efficiency and optimality of the model

The efficiency and optimality concepts are essential requirements for designing, controlling, and exploiting systems. Consequently, it is necessary to offer a definition for these two terms used throughout the book.

In general, as explained in Chapter 2, any state of the area function $A(x)$ of a *tube* of *arbitrary shape* can be perturbed and this perturbation will result in a change of the three formants ($\Delta F_1, \Delta F_2, \Delta F_3$) that could be, but is *not* necessarily, *efficient*. Such change can be made efficient for *one formant*, that is resulting in a maximal ΔF_i with minimal area perturbation, whenever the perturbation is proportional to the sensitivity function *for that formant*. Applying this *efficient* deformation iteratively to a *uniform* tube showed, in Chapter 3, that the perturbation should be exerted in distinctive regions along the uniform tube and, if we exploit a synergy of the front and back regions, this *efficiency* will be *maximized*. Furthermore, the perturbation (i.e., the deformation) may be quite large. It was also shown that the use of this maximum efficiency phenomenon produces *efficient linear trajectories* in the acoustic space.

Optimality in control system design (modeling and control) is seeking the *best possible control* with respect to a performance metric of the *best model* possible (Kwatny, 2000). In our analysis of the optimality of the DRM, we adopted the definition of the basic optimal control problem: Given a dynamic system, find a model and its control that steers the system from an initial state to a target set of states or trajectories while minimizing the cost (in time or energy). When we seek optimal control as a function of the state, the process becomes a closed-loop control. Optimality may contain constraints, such as a specific initial state (state constraint) or some specific control (minimum energy constraint).

One is usually interested in steering a controllable system along a trajectory that is, in some sense, optimal. One of the first methods commonly used to address such tasks is "the calculus of variations" (Browder, 1965), first developed to characterize the dynamic behavior of physical systems governed by some conservation law, which we also discuss in Chapter 2. This method uses Lagrangian systems, one of which is the sensitivity function that we have adopted for the model. A Lagrangian system is defined in terms of a vector of configuration coordinates associated with velocities. The system has kinetic energy *KE* and potential energy *PE*, the difference *KE-PE* of which represents the formal definition of the Lagrangian. The system changes along a trajectory between initial and subsequent states, or between initial and subsequent times, in a way that will minimize the *cost*, which in our case consists of effort to accomplish a certain area deformation that is not only efficient but also exhibits maximum efficiency.

Therefore, by *optimal* DRM we mean that, for the case of the closed-open tube, the various *states* of the system (i.e., deformations of the tube's area function and the specific resonant frequency trajectories associated with them) are determined such that given *criteria of optimality* are met, subject to some given *constraints*.

Our optimality criteria consist of maximum efficiency, maximum acoustic contrast, best number (2^3) and places (regions) for formant transition behavior, and maximum acoustic space, while the constraints include the *uniform tube* as initial state, distinctive regions, and synergy. For the case of the closed-closed tube, some of the optimality criteria and constraints are different (as mentioned in Section 4.2.4).

Efficiency, as defined above, is valid for a *single formant*. When more formants are simultaneously controlled, the system is practically optimal or pseudo-optimal because of the *pseudo-linearity* of the process within the vocalic space and consequently the rule of superposition, or additivity, applies (Broad and Clermont, 2002; Mrayati et al., 1990; Carré and Mrayati, 1990).

4.5 Summary and conclusions

This chapter's contribution can be summarized in several points. *First*, it presents an acoustic model of the tube, derived from an entirely deduction-based self-guided iterative algorithm driven by the principle of efficiency. The model performs time varying efficient deformations of the tube along distinctive spatial regions controllable by transversal commands, incorporates synergy, compensatory and orthogonality aspects of the regions, and displays distinctive dynamic trajectories. It is structured as two, four, or eight distinctive rectangular regions for the closed-open and three or seven for the closed-closed tube. These regions accommodate the results of Chapter 3 and are obtained from efficient use of the acoustics of the tube exploiting sensitivity functions. Each region has its own distinctive formant transition behavior (*FTB*), some of which are efficient only around the uniform neutral tube. This behavior describes in a simple manner the effect of changes in cross-sectional areas of the tube on the formant frequencies. Nomograms relating region area variations to F_1, F_2, and F_3 are derived for an initially neutral tube. Three modes, the One-Tract Mode, the Two-Tract Mode, and the Transition Mode, are recognized. They define (depending on the degree of constriction) when the tube can be treated as a single tightly coupled tract, or when it is actually divided into two tracts, or the transition between these two modes.

Second, a simple control strategy composed of only three commands is formulated, resulting in a *simplified model* derived from the eight-region DRM. This simplified model incorporates both the closed-open and the closed-closed case and its control mechanism and it is well suited to investigating and explaining speech dynamics (see Chapter 7).

Third, using the model with the simplified command, seven distinctive efficient primitive *trajectories* of monotonic formant frequency variations are deduced and defined in the F_1-F_2, F_1-F_3, and F_2-F_3 planes. These trajectories are generated by specific cross-sectional area changes within and across the tube regions. To put it simply, the model is commanded by efficient distinctive primitive area deformations that result in efficient distinctive primitive acoustic trajectories. It is also commanded by constricting specific regions (principally 5 of the 8), that produce efficient *FTB*s (see the nomograms of the eight-region model). This representation entails valuable explanatory and deductive power, as shown in the following chapters.

Fourth, other interesting features of the model, those that no other sound production model to our knowledge possesses and that can explain certain phenomena, are also briefly introduced. They include the anti-symmetrical behavior leading to possible compensatory and synergetic commands as well as the orthogonality, linearity, and optimality of the regions' acoustic behavior; in particular (1) an area increase in one of the front regions could acoustically compensate a similar increase in the corresponding back region, an aspect valid for each of the front four regions; (2) when two corresponding regions change area in an opposite direction, formant changes are synergic since they are additive in the two regions.

Fifth, the DRM results should be viewed in the dynamic domain of time varying "efficient deformations." The distinctive trajectories and the nomograms represent quantitative changes of formant frequencies that reflect their dynamic aspects.

Sixth, when a constriction is located in a given region and a cavity in its symmetrical region, very small variations in the cross-sectional area of a constriction are equivalent (in the acoustic domain) to large variations in its symmetrical large-cross-sectional area region. Nevertheless, if these two variations are presented on a logarithmic scale, they are practically equivalent. This means that, on a logarithmic scale, the degree of narrowing in a region, and the area of its symmetrical region (which is usually a large area), have equal effect. This feature is unique in our model and a consequence of the fact that its regions are symmetrical with respect to the midpoint of the tube (Atal et al., 1978).

Seventh, the DRM model includes several interesting features that could provide simple explanations to acoustic phenomena such as *fibers* – different area functions that yield identical F-patterns – and *formant clamping* (discussed by Schroeder, 1967). In earlier work, we have provided a detailed account for these features (Mrayati et al., 1988).

Finally, in Chapters 2 and 3 our integrated approach resulted in a tube model with distinctive, efficient and optimal characteristics and control strategy. This DRM model – with its distinctive trajectories, its regions related to places of distinctive formant behavior, and its nomograms – can be a suitable device for sound production. Since the following chapters present ample evidence showing analogy of the model and the natural speech production system, it is not unrealistic to propose that, having discovered such a tube at their disposal, human beings have made use of it as an effective communication tool. As presented in Chapters 5, 6, 7, and 8, the model enables us to predict vowel systems, production of vowel and stop-consonant sequences, as well as the dynamics of speech production and perception.

5 Speech production and the model

In Chapters 3 and 4 we presented *tools* for an acoustic production system. This chapter compares characteristics of human speech production and those of the model. The chapter demonstrates that data generated by the model are in remarkably good agreement with speech production results: both display maximum formant spaces outlining the same vowel triangle, show identical gestural deformations, identical synergetic deformation of the tube, and identical places of articulation for vowel and consonant production. The good agreement implies that oral vowels and plosive consonants are intrinsic physical properties of the acoustic tube. This chapter takes the formant transitions and the acoustic trajectory primitives produced by the model, places them in a syllabic frame, and generates vowels as dynamic trajectory states rather than static targets. It also provides a deductive explanation of human speech production and presents it as the consequence of specific efficient deformations of the tube, performed with the aim of achieving maximum acoustic variations.

In the system based on an acoustic tube (see Chapter 3), we deduced a set of efficient area deformations and their corresponding trajectories inside the acoustic space in which they operate (specifically, the F_1-F_2 plane). These results were represented by the Distinctive Region Model (DRM) outlined in Chapter 4, a model that divided the tube into eight distinctive regions.

To reach our objective of discovering how the tube model explains the production of speech by the vocal tract, we begin by using a simplified form of the eight-region model. Taking the first two formants into account, for the closed-open model type, the acoustic tube is divided into four (Figure 5.1 a), and for the closed-closed type into three regions (Figure 5.1 b). Due to the anti-symmetrical behavior of the closed-open case, the model has two controls to perform, a front and a back constriction (with the corresponding cavities) and a constriction at the tube's aperture. With the symmetrical behavior of the closed-closed case a central constriction can be performed. When taking into account three formants, the closed-open tube is divided into eight regions (Figure 5.1 c). In this case, the nomograms of all three formants of the eight regions show orthogonal acoustic behavior (around the neutral configuration, see Figure 5.2 a). In contrast, the closed-closed tube is divided into seven regions (Figure 5.1 d). In this case, F_1 is equal to the vibration frequency of the walls (that, for all practical purposes, does not change), and the pseudo-orthogonal appearance disappears (Figure 5.2 b).

To create a realistic model of the vocal tract – and for reasons of computational ease – the *simplified eight-region model* (the *DRM*) described in Subsection 4.3.1 will be used. When only F_1 and F_2 are taken into consideration, this

DOI 10.1515/9781501502019-006

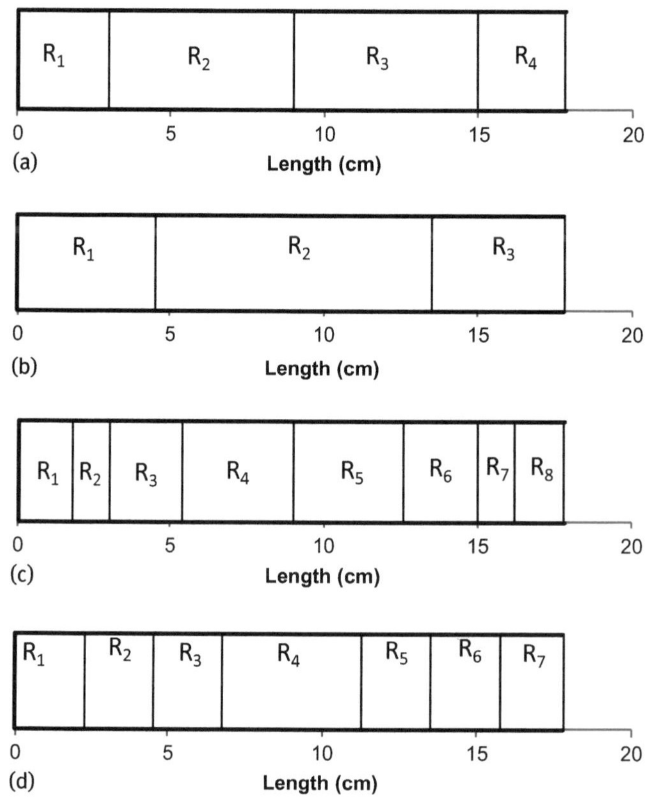

Fig. 5.1: Panel (a): Anti-symmetrical four-region Distinctive Region Model (DRM) used for a closed-open tube taking into account F_1 and F_2. Panel (b): Symmetrical three-region DRM, used for the closed-closed tube taking into account F_1 and F_2. Panel (c): Anti-symmetrical eight-region DRM used for a closed-open tube taking into account F_1, F_2, and F_3. Panel (d): Symmetrical seven-region DRM used for the closed-closed tube taking into account F_1, F_2, and F_3.

simplified model is able to represent: (1) a closed-open four-region model (Figure 5.3 a); (2) a closed-closed three-region model (Figure 5.3 b). In this model, regions R_4 and R_5 in the center of the tube generate the constriction approximating the central constriction of the theoretical model (R_2 of Figure 5.1 b). When taking into consideration F_1, F_2 and F_3, such a model is also able to represent an eight-region model with each region having a specific formant transition behavior, as shown in the nomogram cartoon of Figure 5.2 a. Region R_1 is fixed and corresponds to the source cavity, while Region R_8 corresponds to the output aperture (panels a and b of Figure 5.3).

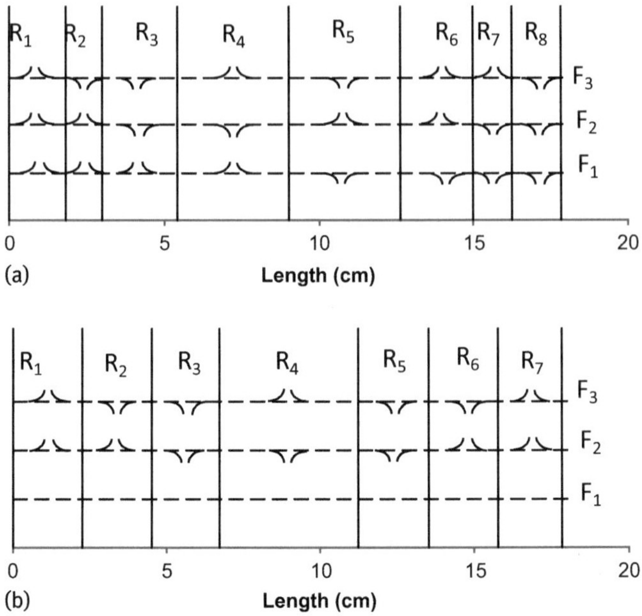

Fig. 5.2: Panel (a): For the closed-open tube, schematized eight formant transitions when closing and opening the eight regions starting from a neutral (i.e., uniform) configuration. Panel (b): For the closed-closed tube, schematized seven formant transitions when closing and opening the seven regions starting from a neutral configuration.

This simplified model covers both the closed-open and the closed-closed structures; it functions with two main commands that act either alone or jointly (the first of which achieves front, back, or central contriction by transversal or longitudinal actions, as explained in Subsection 3.2.3.3, and the second activates the aperture) and leads to the primitive F_1-F_2 trajectories presented in Subsection 4.3.3 (Figure 5.3 c).

In this chapter, we will first compare the DRM regions to the main regions of the vocal tract as well as to the main places of articulation observed during the production of oral vowels and plosive consonants. Then, using the essentially dynamic approach we developed, the trajectories in the F_1-F_2 plane (Figure 5.3 c) will be examined. These trajectories are quasi-rectilinear and they derive from either the activity of a single deformation gesture (that of the "tongue" with front, central, and back place of constriction) or from the co-production of two gestures (those of the "tongue" and the "lips"). By co-production we mean that the two gestures are produced simultaneously. In an acoustic communication system the presence of an independent second gesture doubles the set of trajectories and, in

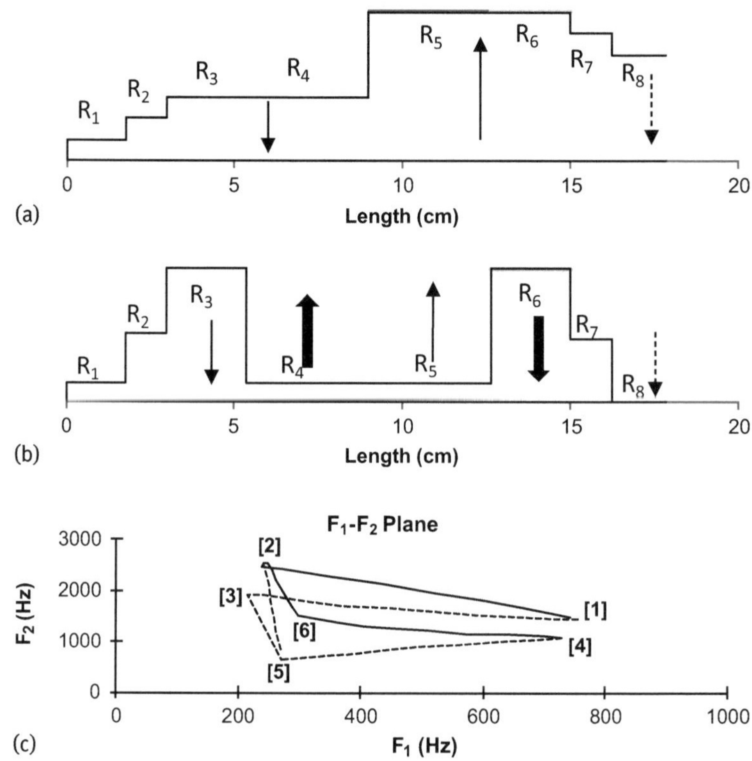

Fig. 5.3: The simplified eight-region model. Panel (a): a closed-open four-region model controlling both F_1 and F_2; panel (b): a closed-closed region model; (c) the primitive trajectories in the F_1-F_2 plane of the simplified model.

theory, it increases the command code space by a factor of two (by way of maximal use of a new code unit – or maximal use of a new distinctive feature (Ohala, 1979). We will observe that the main oral vowels can be located on these primitive trajectories.

By deduction, in the DRM's nomograms the alignment of the three resonant frequencies with the area functions (Subsection 4.2.3 and Figure 5.2) leads to acoustically distinctive occlusive sounds generated by closing and re-opening any of the eight distinctive regions of the acoustic tube. We will show that these regions correspond to the main places of articulation for producing plosive consonants.

As in some cases vowels are co-produced by two gestures, a CV-syllable will also be considered as the co-production of gestures, that is, parallel production of information pertaining to the vowel and/or the consonant. Since co-produced gestures in V_1V_2 (vowel–vowel) and CV represent primitive elements of the communication code, they must enjoy a certain degree of autonomy. If they were always co-produced under identical conditions with identical characteristics, they would be reduced to a single code element. A general scheme of syllabic production will be proposed.

In summary, using a deductive approach allows us to infer the known standard places of articulation for vowels and consonants and thereby make it possible to identify primary physical underpinnings of the most important phonological distinctions.

5.1 The articulatory level

The cartoon shown in Figure 5.4 maps the eight regions of the model into the vocal tract for the closed-open tube when taking into account three formants. As the figure indicates and as is explained in Chapter 3, the total effective length L of the tube includes the radiation inductance effect L_r. The figure illustrates several important features:

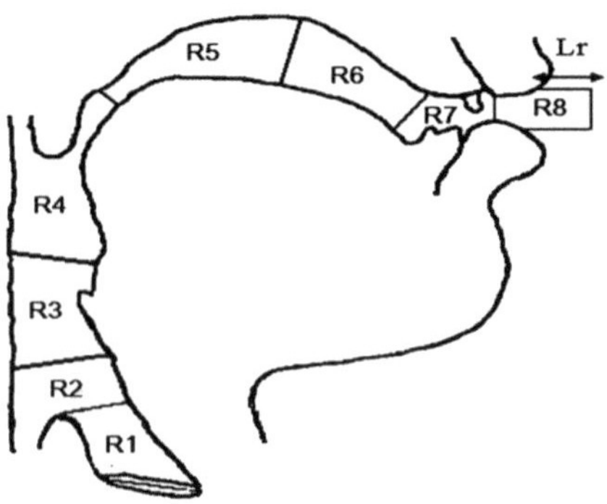

Fig. 5.4: The eight DRM regions (closed-open tube) mapped in the vocal tract. R_1 corresponds to the larynx cavity, R_3, R_4, R_5, R_6 to the tongue, R_7 to the teeth, and R_8 to the lips plus the radiation effect L_r.

- The model's regions are aligned with the *anatomy* of the articulators: Region 8 (with the radiation effect) corresponds to the lips, region 7 to the teeth, regions 3, 4, 5, 6 to the tongue, and region 1 to the larynx. Regions 2 and 7 can be considered as intermediate regions the acoustic effects of which on the formant are small (see Chapter 4).
- The *synergy* of the system observed in Chapter 3 and modeled in Chapter 4 is achieved by virtue of the *constant volume* of the tongue: The automatic deformations, anti-symmetrical for the case of a closed-open tube (Figures 3.10 a and 3.10 b) and symmetrical for a closed-closed tube (Figure 3.14 a), naturally

occur in the human vocal tract (see Chapter 1): a front constriction gives rise to a back cavity and vice versa, while a central constriction automatically generates both front and back cavities. The absolute size of the cavity (in cm^2) is acoustically less important than the absolute degree of the corresponding constriction (on a logarithmic scale, constriction or cavity areas have the same acoustic effect, see Chapters 2 and 3). This means that the constant tongue volume carries a crude representation of this synergetic effect. Thus, the control of the anti-symmetry/symmetry mechanism is much more sensitive to the size of the constriction than to the size of the corresponding cavity. This is naturally achieved because the tongue in the human tract is situated at the middle of the vocal tract and it is thus able to exploit the anti-symmetry and symmetry as well as the synergy properties of the tube.
- As a consequence of the anti-symmetrical acoustic behavior of the closed-open tube, Regions 3 and 4 will exhibit features similar to those of the back (pharyngeal) cavity and regions 5 and 6 will exhibit features similar to those of the front (mouth) cavity. Thus, a *pharynx cavity* is automatically obtained when a front constriction occurs. As a consequence of the symmetrical acoustic behavior of a closed-closed tube, regions 4 and 5 in the model will exhibit features opposite to those of the back and front cavity (regions 3 and 6).
- The *area deformations* corresponding to maximum formant changes, as postulated in Chapter 3 and modeled in Chapter 4 (see the *OTM*, Subsection 4.2.3.3), have a *range* between 0.5 and 10-to-16 cm^2. This range is similar (and quasi-identical) to area functions seen in Fant's X-ray measurements of vowels produced by talkers speaking at a moderate vocal intensity (Fant, 1960).

Consistent with Fant's measurements, the three main places of articulation in vowel production corresponding to [i], [a], and [u] in the corners of the vowel triangle also correspond to the corners of the triangular space determined in the acoustic tube by iterative calculation (Chapter 3, Subsection 3.2.1). The area functions of these three vowels can be obtained with a four-region model controlling F_1 and F_2 (Chapter 4, Subsection 4.3.3). The closed-open model predicts, and coincides with, front (R_5, R_6) and back (R_3, R_4) constrictions, whereas the closed-closed model predicts the central constriction (approximately R_4 and R_5, Figure 5.5). In the human articulatory system these constrictions are performed by the three main tongue muscles (genioglossus, hioglossus, and styloglossus).

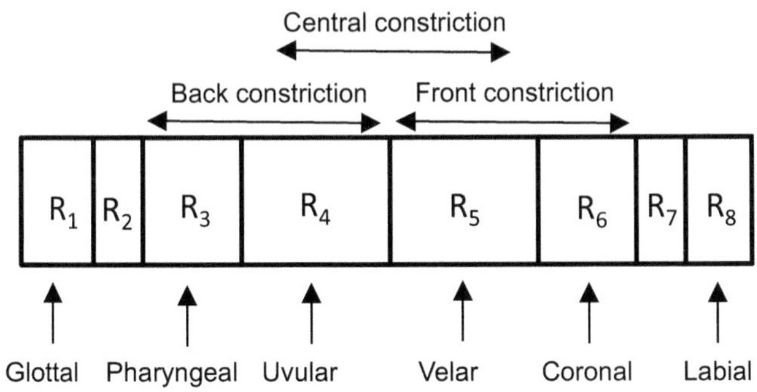

Fig. 5.5: The eight DRM regions and the vowel and consonant places of articulation.

Regarding consonant production, two points need to be raised:

- Considering the ensemble of the majority of languages, six existing main *places of articulation* of consonant production will be considered: they correspond to regions of an eight-region model (labial: R_8, coronal: R_6, velar: R_5, uvular: R_4, pharyngeal: R_3 and glottal: R_1), see Figure 5.5. This model is more complex than the one for vowel production; for example, for the distinction between velar (R_5) and coronal (R_6) consonants, the vowel front constriction must be divided into two regions. The distinctive regions of the model offer optimal places for consonant articulation with the areas changing between 0 and 16 cm^2 (see Figure 5.2 for FTBs of these regions).
- To make the six consonants generated by constricting each of these regions acoustically distinct, the third formant must also be taken into account (Carré and Mody, 1997).

Observations of vocal tract area function changes during speech production show that a large proportion of the deformations is *transversal* and a smaller proportion *longitudinal* (see, for example, Iskarous, 2001). As described Chapters 3 and 4 (Sections 3.2.3.3. and 4.3.2), longitudinal displacement of a constriction is obtained during the passage from a closed-open to a closed-closed model. In the articulatory system, the three main "tongue gestures" consist of a transversal front to back displacement of the tongue constriction (and vice-versa) and two longitudinal displacements from front to center and back to center (and, naturally, vice-versa). These three gestures are similar to those proposed by Gunnilstam (1974) based on articulatory data for vowel production.

5.2 The DRM and vowel production

The deductive approach used to study the properties of an acoustic tube leads to the following points to consider:

- The tube defines an acoustic space similar to the vowel triangle (Figure 3.15); the size of the vowel triangle is at its maximum and it cannot be further increased. Obviously, the model also allows for working with a vowel space that can use *sufficient* contrasts.
- In the model, constriction displacements together with the aperture control correspond to the three main tongue gestures mentioned above plus changing the lip opening. In natural speech the three tongue gestures and the lip gesture together produce formant transitions along specific trajectories. But we have shown in the model described in Chapter 4 that the constriction displacements and aperture change produce seven distinctive trajectories in the F_1-F_2 plane. What we want to investigate now is how these model trajectories map into those produced by the above four gestures in real speech (Chapter 4 and Figure 5.3 c).
- These area function/articulatory gestures and the acoustic primitive *trajectories* that they generate can be regarded as constituting the *coding units* for the vowel space's structure. In Figure 5.3 c the solid lines represent trajectories produced by the three tongue gestures and the broken line the trajectories co-produced by the tongue gestures and the lip gesture. The figure clearly shows that the number of trajectories generated by the co-produced gestures is identical to those produced by the simple gestures. Therefore, co-production of gestures increases the code repertoire by a factor of two. The seventh trajectory [2]-[5] in the figure is a linkage between the two triangles (solid line and broken line). The DRM represents a production system akin to parallel computing – two sets of information are co-produced in parallel and the transmission capacity of the acoustic communication system can be doubled.

The primitive trajectories obtained by deduction and the vowels proposed by phoneticians (see, e.g., Catford, 1988) are represented in the F_1-F_2 plane in Figure 5.6. This figure illustrates the trajectories with 18 vowels positioned in the plane (Maddieson, 1984) and identified by their corresponding phonetic symbols. The vowels lie on the trajectories. To obtain additional vowels, one or more additional gestures is/are needed (e.g., nasal, advanced tongue root [ATR]). Furthermore, the main diphthongs listed in language inventories (e.g., Maddieson, 1984) also lie on the primitive trajectories.

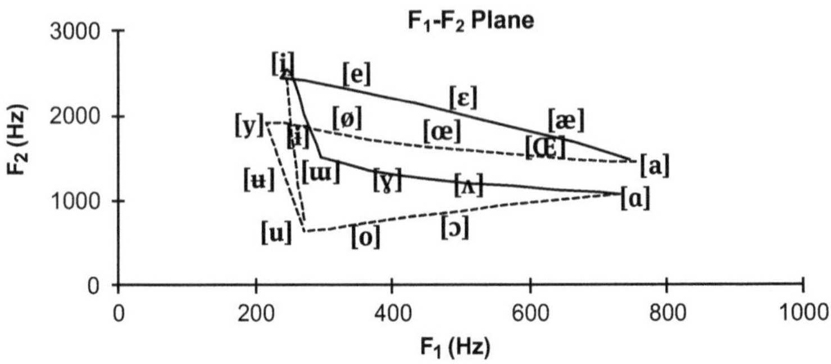

Fig. 5.6: Formant trajectories in the F_1-F_2 plane obtained automatically are the basic dynamic coding units of the speech communication system. The vowels are sub-products.

Since the trajectories are dynamic and the vowels lie on them, the vowels themselves should be viewed as consequences of a dynamic process rather than as static targets: they can be considered byproducts of vowel trajectories.

Figure 5.6 shows the phonetic and phonological potential of the acoustic tube that undergoes efficient and simple deformations, a potential that derives from *trajectories*, rather than from static positions. Each formant trajectory is obtained using the model to generate a single *primitive* gesture or two co-produced primitive gestures – one corresponding to the tongue constriction displacement (associated with cavities) and the other to the lip opening.

To summarize, the DRM has produced three main trajectories, [ai], [au], and [iu] and four complementary ones: [ay] labialized, [aɯ] and [iɯ] non-labialized, and [uy] fully labialized. Our approach emphasizes that formant trajectories produced by two elementary speech gestures ("tongue gesture and lip gesture") are key components of the speech communication system we are about to build. Vowels lying on the trajectories are direct consequences of these trajectories. This approach stresses the *dynamic* aspect of vowels, although they are frequently described as static events.

5.3 Vowel dynamics

If vowels are byproducts of primitive vocal trajectories, it means that their actual positions on the trajectories may not be stable. This hypothesis was first tested using the DRM to produce the 11 standard oral French vowels (Carré and Mrayati, 1990, p. 231, their Figure 19). To examine the acoustic stability of these vowels, for each vowel the areas of each section of the eight-region model were successively multiplied or divided by 1.41; as Figure 5.7 indicates, each of these perturbations of the area function changed the values of the first two formants.

The formant variations for the vowels [i], [e], [ɛ] lie clearly on the [ai] trajectory, while those for the vowels [u], [o], [ɔ], [ɑ] are on the [ɑu] trajectory. For [ai] this result confirms that the acoustic effect of labialization is quite small when starting from an initial state of fully open lips. The vowel that is supposed to be on the trajectory [ay] is unstable because it results from two independent gestures having orthogonal acoustic effects and the acoustic effect of a small lip aperture (Region 8 of the DRM) is large.

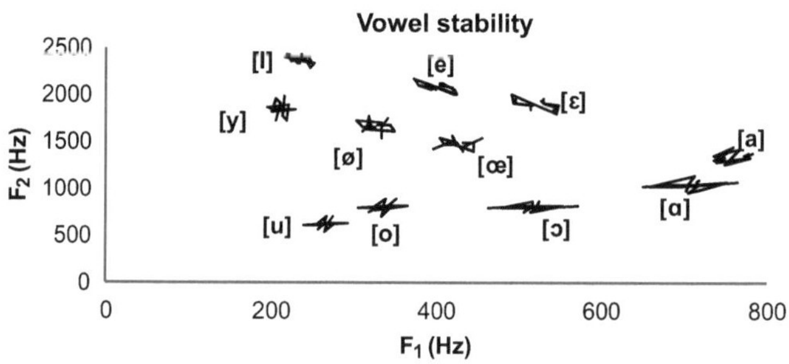

Fig. 5.7: Formant frequency variations obtained when the cross-sectional area of each of the eight regions is multiplied or divided by a factor of 1.41. The 8-region configurations correspond to 11 French vowels, (after Carré and Mrayati, 1990, p. 231, Figure 19, p. 231, with permission by Springer Publishing Co.).

In natural speech, this dynamic aspect of vowel production can explain the instability, the "vowel-inherent spectral change," in isolated vowels, as well in the vowel of some natural CVCs (Nearey and Assmann, 1986; Morrison and Assmann, 2013). In fact, this change appears to follow the underlying primitive vocalic trajectories (see Figure 5.8). A similar phenomenon can also be observed in French V_1V_2 production (see Carré and Mrayati, 1991, their Figure 1).

Synchronization of the two gestures appears to have an important role for the perception of two central vowels that may have approximately identical formant frequencies (for example, /ɯ/ and /ø/) but very different production characteristics: the first is obtained by a central tongue constriction without labialization and the second by a front tongue constriction with labialization. Both vowels can be present in some languages, like Bedjonde (Djarangar, 1989). Although the formants in the F_1-F_2 plane can be more or less identical for *these* two vowels, trajectories leading to these acoustically identical vowels from a third vowel are different. This situation highlights the *dynamic* aspect of vowel production, as represented in Figure 5.9. As an example, in DRM a longitudinal tongue gesture is used

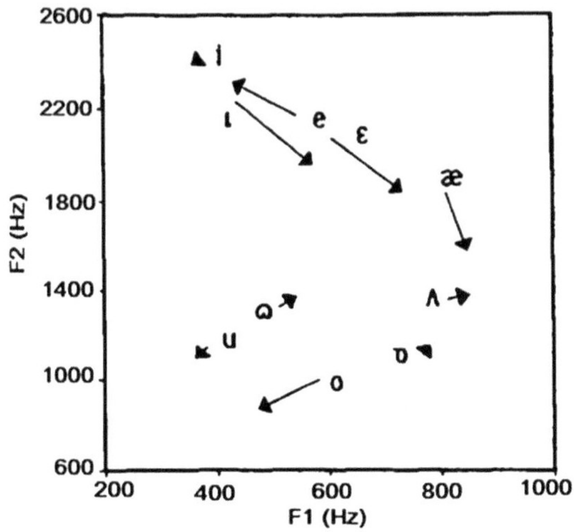

Fig. 5.8: Inherent spectral changes of isolated vowels (from Nearey and Assmann, 1986, Figure 2, p. 1300, with permission by the Acoustical Society of America).

to produce [iɯ] resulting in an approximately rectilinear trajectory, while two gestures – a transverse tongue gesture and a synchronized lip gesture – are required to generate [iø] and produce a curved trajectory (Figure 5.9 a). The curvature of the trajectory is the result of the two co-produced gestures; the acoustic effect of the lip gesture follows that of the constriction gesture after a delay. If we made the two gestures asynchronous by advancing the onset of the lip gesture, the trajectories would become rectilinear again. The question here is whether two different trajectories between the same two vowels will exhibit an acoustic contrast sufficient for the different trajectories to be also perceptually different – e.g., whether the lip rounding is perceived in the production of [ø].

Vowel trajectories derived by the DRM represent primitives of efficient coding and predict the use of the same trajectories in the chain of vowels found in natural languages. According to Öhman (1966a), speech represents a flow of vowels on which consonants are super-imposed. As a test of this idea, we tabulated the vowel succession in the Sankoff–Cedergren corpus of spoken Montreal French (Sankoff et al., 1973). The procedure consisted of tabulating the number of occurrence for each V_1V_2 succession while disregarding all consonants. Results of this tabulation, illustrated in Table 5.1 (Carré et al., 1995), show that the most frequent vowel successions were those situated on the trajectory between [a] and [i], such as [aε], [ea], [eε], etc., and their permutations. From the standpoint of coding (as already explained in Chapter 3), vowels on the same trajectory represent a gesture with different degrees of constriction (and also different velocities of change, as elaborated in Chapter 7). Consequently, the preponderance of vowel successions on the [a]-to-[i] trajectory suggests a preference of Montreal French for efficient and simple coding. Conversely, successions of vowels on different trajectories increase the contrast (and thus the distance on some perceptual scale) at the cost of decreased efficiency and simplicity. Since the frequency of such successions is much lower than those of vowels on the same trajectory, one must conclude that a language exhibits a clear preference for the economy provided by efficient,

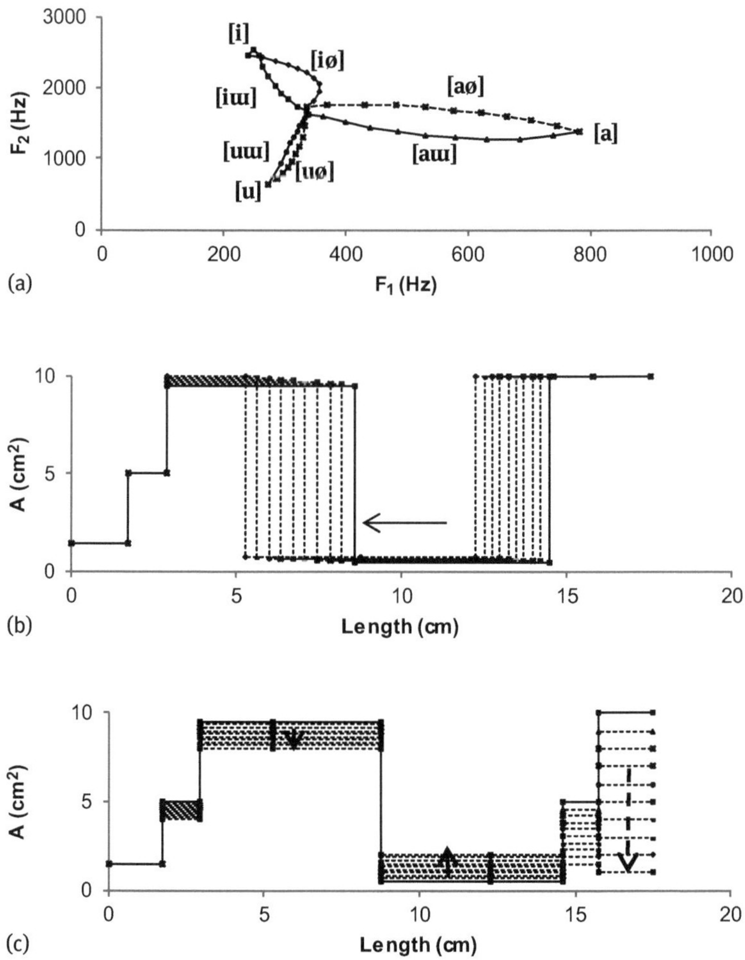

Fig. 5.9: Panel (a): Trajectories [Vɯ] and [Vø] generated by the DRM model in the direction of two acoustically identical vowels: One is obtained with central constriction and no labialization, and the other with front constriction and labialization. V= [a], [i], or [u]. Panel (b): From front constriction [i] to central constriction [ɯ]. Panel (c): From [i] to [ø] with labialization and change in the degree of constriction.

Table 5.1: Number of successions of V_1V_2 in the spoken corpus (the Sankoff–Cedergren corpus of Montreal French; Sankoff and Sankoff, 1973) disregarding all intervocalic consonants (based on Carré, Bourdeau & Tubach 1995, Table 3, p. 210, with permission by Karger Publishers).

[aɛ]:	8515	[ea]:	5525	[eɛ]:	5068	[ii]:	4978	[ee]:	4686
[ia]:	4667	[ae]:	4612	[aa]:	4584	[ie]:	4459	[ai]:	4283
[ɛe]:	4120	[ɛa]:	3833	[ei]:	3766	[ɛɑ]:	3361	[ɑi]:	3296
[ɑe]:	3119	[eɑ]:	3018	[iɔ̃]:	3011	[ɑ̃a]:	2907	[ɑ̃e]:	2892
[iɑ]:	2867	[ɑa]:	2849	[ɛi]:	2749	[eɑ̃]:	2558	[ɛə]:	2542
[iɑ̃]:	2443	[ɛɑ̃]:	2291	[aɑ̃]:	2280	[əa]:	2268	[iɛ]:	2234
[ɛɛ]:	2194	[aə]:	2180	[ɔ̃a]:	2107	[ɑ̃i]:	2093	[ɔ̃e]:	1992
[əɛ]:	1973	[ɑ̃ɑ̃]:	1965	[əi]:	1899	[ey]:	1816	[iə]:	1761
[ye]:	1759	[əe]:	1757	[eu]:	1709	[ɑ̃ɛ]:	1676	[aɑ]:	1675
[ɑ̃ɑ]:	1653	[ya]:	1634	[ue]:	1627	[eɔ̃]:	1614	[ɛy]:	1600
[aɔ]:	1571	[ɑɑ]:	1501	[ua]:	1497	[ui]:	1438	[əə]:	1392
[aɔ̃]:	1377	[ɛɔ̃]:	1377	[ɔɛ]:	1368	[əɑ̃]:	1336	[ɑɛ]:	1301
[ɔ̃ɑ]:	1277	[ɑ̃ə]:	1273	[ɔ̃ɛ]:	1264	[ɔ̃i]:	1254	[ɔi]:	1245
[ou]:	1219	[ɔe]:	1151	[uɛ]:	1147	[ay]:	1138	[eə]:	1122
[iu]:	1121	[ɛu]:	1120	[ɑɔ̃]:	1108	[oi]:	1104	[eo]:	1100
[ɛ̃e]:	1099	[yi]:	1065	[ɔa]:	1053	[ɑə]:	1050	[ɑɑ̃]:	1048
[ɛ̃i]:	1027	[ɑy]:	1005	[ɛɛ̃]:	999	[ɔ̃ɑ̃]:	976	[ɑ̃ɔ̃]:	955

simple coding (least effort principle). Nasality is a specific case but, although it is represented in the table, it will not be discussed here.

5.4 Consonant production and the model

In our approach, consonants and vowels are produced within the same theoretical framework. The regions and trajectories of the model are the essential elements of the production of consonant places of articulation and the formant transition pattern of vowels. It is important to stress that within the vocalic space the formant frequencies are linearly related to the area function as long as one stays inside the 0.5 to 10–16 cm^2 range (identified as One Tract Mode, *OTM*, in Chapter 4). Conversely, the relationship becomes non-linear for consonant occlusions producing areas smaller than 0.5 cm^2 (called Transition Mode, *TM*, and Two Tract Mode, *TTM*, in Chapter 4).

5.4.1 CV syllabic co-production and the model

Distinctive vocalic and consonantal gestures have been deductively obtained from the acoustic tube (Chapter 3) and substantiated by the DRM (Chapter 4). The ques-

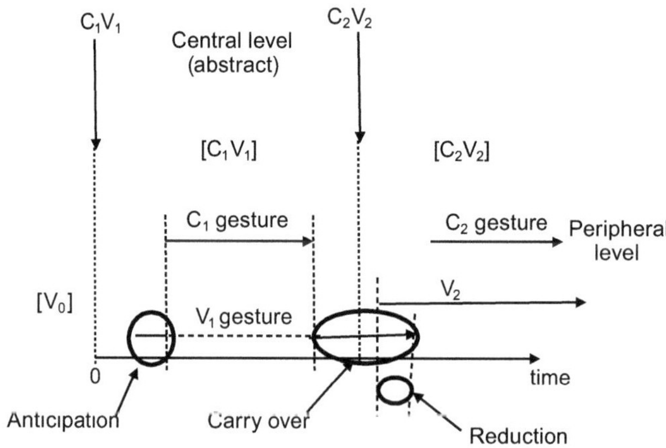

Fig. 5.10: Timing sequence of syllabic co-production of vowel and consonant gestures, indicating command (central level) and execution (peripheral level). As a function of the gestures' relative timing, anticipation, carryover, and reduction can be observed. The dashed portion of the vocalic gesture line indicates that the vocalic gesture is stopped by the consonantal gesture; it recovers during the carryover period but it is reduced for a period after the onset of the second vowel.

tion is, *how these gestures are used* to produce speech. Because we are dealing with a dynamic process, CV syllables could be produced either by *temporal succession* (commands and gestures successive without overlap), by *coarticulation* (although the commands of C and V are successive, the gestures overlap, and the CV syllable is produced with overlapping but desynchronized C and V gestures), or by *co-production* (commands of C and V are simultaneous and the gestures overlap). The consonant gestures and the vocalic gestures of the CV syllable are produced more or less simultaneously – in *parallel* (increasing the efficiency of coding). But, because the gestures are distinctive, they are more or less independent. This means that their duration, their synchrony, and their time function have a certain degree of freedom. These points will be discussed more fully in Chapter 7.

As shown in the scheme represented in Figure 5.10, starting at the initial configuration V_0, the syllabic command to execute the gestures C_1V_1 is issued at some central level and it is sent to the peripheral level at time 0. Each separate speech gesture is characterized by its type (vocalic or consonantal gesture), duration, kinetic function, and onset/offset time on a time scale starting at 0. Although the command is abstract and discrete, many of its characteristics are continuous and reflect physiological, linguistic, sociological, and behavioral parameters proper to a given individual.

If the vocalic gesture corresponding to the production of V_1 (two gestures may be involved: tongue and/or lip gestures, at various degrees of synchrony) begins before the consonant gesture C_1 (in an act of anticipatory coarticulation), then the formant transition at the end of V_0 has a V_1 bias. If the C_1 gesture terminates before the end of V_1 (termed the carryover effect), the formant transitions going into V_1 are influenced by V_0. Furthermore, if a new syllabic command to execute the gesture V_2 (or C_2) arrives before the end of the V_1 gesture, it will prevent the V_1 gesture from being fully executed (termed the reduction effect).

Where is the beginning of the syllable in such a timing scheme? If it were at time 0 at the onset of the abstract syllabic command at the central level, it would be impossible to observe. Therefore, the observable syllable must start at the peripheral level, that is, in the example of the virtual talker given in Figure 5.10, at the onset of the vocalic gesture rather than that of the consonant. A different virtual talker could start the syllable with the onset of the consonant.

To define what is meant by co-production in the command of the model, we need to distinguish the different interpretations proposed by other approaches using the coarticulation term. In natural speech, coarticulation effects have been extensively investigated (e.g., Farnetani and Recasens, 1999; Kent and Minifie, 1977). In V_1CV_2 production, the onset and offset of the consonant's formant transition often contains information on V_1, C, and V_2. To explain "coarticulation," Hockett (1955) compared the phonemes representing a word to Easter eggs and speech to an omelet obtained with these eggs – that is, the phonemes cannot be separated. Harris et al. (1958) observed that the signal corresponding to "w" extracted from "wip," when spliced together with the signal representing the word "up," is never perceived as "wup." The spectral representation of speech shows the reciprocal influences of a phoneme on the preceding and following ones. Fant (1973) illustrated the coarticulatory effects by extracting time waveforms from a series of phonemes representing a sentence, and found that segmenting the speech signal into phonemes was not possible because of the reciprocal effect of the individual phoneme waveforms. Such effects produce formant transitions that depend on the context and lead to vowel and consonant reduction phenomena. Looking at these results, the questions arise: What is the size of the coarticulation unit that represents all the reciprocal effects observed? And are the coarticulatory effects observed in the speech signal planned at the central level?

Furthermore, in V_1CV_2 production Öhman (1966b) noted traces of V_2 on the acoustics of V_1. This finding suggests that the consonant gesture is superimposed on a vowel-to-vowel gesture and that these two gestures are actually co-produced. In fact, when Kozhevnikov and Chistovich (1965) propose that the main unit of production is the syllable, they base their proposition on the assertion that co-produced consonantal and the vocalic gestures are synchronized. However, this proposition does not take account of the previously cited case by Öhman and the case of consonant clusters between vowels (V_1CCCV_2) where the onsets of the consonantal and the vocalic gestures cannot be synchronized (because in a $V_1C_sV_2$ syllable the lip protrusion needed for V_2 production can begin on V_1, see Benguerel and Cowan, 1974). But, regardless of the synchronization issue, a syllabic gestural co-production mechanism is able to account for articulatory phenomena all along a V_1CV_2 sequence, provided that the vocalic gesture in CV_2 precedes the consonant gesture. Therefore, it is not necessary to propose a specific model for V_1CV_2 (Daniloff and Hammarberg, 1973). Rather, assuming CV co-production leads to a far simpler solution and allows a C_1V_1 command to absorb a host of generally observed coarticulatory characteristics starting at the initial state V_0. Here, gestures are not as they are understood in articulatory phonology. Instead, they are DRM tube deformations directly linked to distinctive acoustic trajectories. These DRM deformations constitute the primitives of a phonological system that correspond to a coordinated set of the various articulatory parameters used in articulatory phonology.

Co-production of articulatory gestures has also been adopted by proponents of articulatory phonology (Browman and Goldstein, 1992). When several gestures are produced to obtain a phonological goal, the relative timing between them is speaker- and language-dependent. In particular, for the timing between consonantal and vocalic gestures in CV_2, anticipation of V_2 at the end of V_1 may be, but not necessarily is, observable in the acoustics. Such an approach is able to provide an explanation for the seemingly contradictory coarticulatory results between Swedish and American English, on the one hand, and Russian, on the other (Öhman, 1966a). A new syllabic command to produce gesture for V_2 (or C_2) before the V_1 gesture has terminated leads to a vowel reduction, as observed by Lindblom (1963).

In a co-produced gesture mechanism, a consonant gesture can interrupt a vocalic gesture (e.g., the C_1 gesture interrupts the V_1 gesture, as shown in Figure 5.10). Some authors (e.g., Gick and Wilson, 2006) speak of "articulatory conflicts." But is there a real conflict? In an orthogonal and even a pseudo-orthogonal transformation of articulatory gestures to acoustic trajectories, an "articulatory conflict" would necessarily have to lead to an "acoustic conflict," which is a meaningless notion when one considers a signal: in physics, a signal or any instance of a signal is either present with all its attributes or it does not exist. This point should merit further investigation. It is worth noting that, although in our approach the model commands are centrally issued, it is unnecessary for all parameters of coarticulation to be planned at this central level. Nevertheless, VCV production by the DRM can be used to test the co-production model proposed above, as shown in the following section.

5.4.2 VCV production

The production of VCV utterances using the specific commands of the DRM and the co-production principle described in the previous subsection is simple and consistent with natural speech results (Carré and Chennoukh, 1995). We took the model to reproduce the set of V_1CV_2 utterances with [b, d, g] as consonants (produced by closing and opening regions R_8, R_6, or R_5) and used the vowels ([y, ø, ɑ, o, u]) Öhman (1966a) studied to illustrate the principle that he named *superimposition* (and that Fowler, 1980b, called co-production). In our simulations, the gesture generating the transition from the first to the second vowel and the closure gesture of the consonant are simultaneous and have identical duration (Figure 5.11). Actually, this is a special case of the general syllabic command illustrated in Figure 5.10. As an example, Figure 5.12 shows the effects of superimposing a consonant gesture on a vowel-to-vowel transition for the production of [ɑby] by the DRM ([b] is obtained by closing and opening region R_8). The formant transition at the end of [ɑ] already manifests the influence of [y], and vice versa. Comparison of [ɑby] from Öhman's data with the simulation by the model is illustrated in Figure 5.12; it indicates a good agreement.

The model successfully reproduces the transition directions for the first, second, and third formants in the data. The good data-model agreement could be mainly attributed to the DRM carefully selecting the relative timing between the tongue and the lip gestures producing the V_1CV_2. It is known that the timing patterns are speaker-dependent. Since timing schemes are not fixed in the model, the DRM together with the adopted co-production principles could be made to produce utterances replicating individual timing characteristics of any speaker.

For pharyngeal plosive consonants some work has also been done using the DRM (see Al Dakkak et al., 1994).

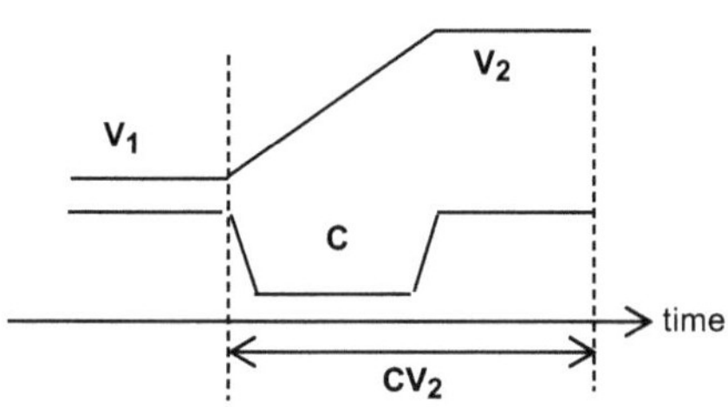

Fig. 5.11: In V_1CV_2 co-production, the consonant gesture is superimposed on the vowel-vowel transition.

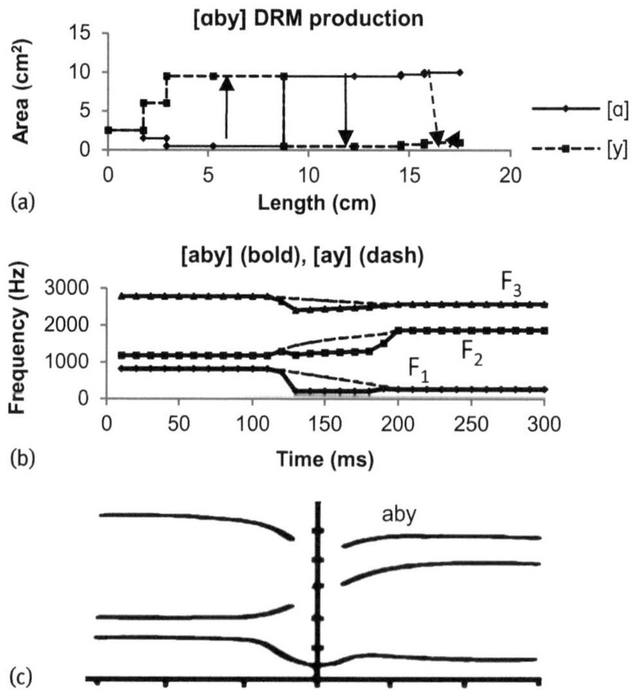

Fig. 5.12: Superimposition of a [b] closure on a [ɑy] transition. Panel (a) shows the vowel and consonant DRM gestures going to [by]. Panel (b) shows the effects of the superimposition on the formants from [ɑ] to [i], both for the case of [ay] (dash) and for that of [aby] (bold). Panel (c) shows the corresponding formant transitions obtained by Öhman (1966, Figure 7, p. 160, with permission by the Acoustical Society of America).

5.5 Discussion and conclusions

This chapter shows that the DRM is able to produce synthetized analogs of natural speech: it acts similar to the vocal tract and, therefore, it can be considered to represent the production-half of a realistic speech communication system. Formant transitions naturally deriving from the model overlap with vocalic formant transitions found in essentially all languages and the maximum acoustic space defined by the model is the vowel triangle. In the vocal tract, the model identifies, instead of areas of quantal *stability* as proposed by Stevens (1989), distinctive regions that serve as anchors for acoustic *changes* (specific formant transition behavior, FTB) and that correspond to the places of articulation of vowels and consonants.

The fact that vowels emerge from formant *trajectories* rather than quantal targets brings into focus the importance of the *dynamic* nature of vowel production:

they are not targets and they are not intrinsically stable. Rather, they are byproducts of vocalic trajectories.

Regions of the model deductively obtained from the tube correspond to natural places of articulation. In fact, both vowels and plosive consonants are derived using the *same theoretical framework* – a conclusion also proposed by Clements (1993) on phonological grounds.

Results obtained by our approach, being in good agreement with empirical data, lead to the following predictions:

- Since consonantal places of articulation are predicted by the model taking into account the first three formants, we are led to postulate that the *bursts* associated with plosives (Blumstein and Stevens, 1979; Stevens and Blumstein, 1981) are a consequence of the primary objective – control of the transitions' formant frequencies. Indeed, it has been shown that formant transitions alone are sufficient for plosive identification (Dorman et al., 1977). This explanation does not mean that stop bursts are unimportant for perception: the burst generated by the opening of the closure at a given place of articulation has its own spectral characteristics that can provide a perceptual cue in addition to the transition.
- The DRM predicts that the third formant should be essential for differentiating consonants, like /d/ and /g/. In fact, Harris et al. (1958) noted that the consonant /d/ cannot be obtained in the vocalic context /i/ without the third formant being present. In addition, using acoustic V_1CV_2 data from Öhman (1966a), Lindblom (1996) showed that for a CV syllable the locus of /b/ exhibits no overlap with those of /d/ and /g/, whereas, on the basis of F_2 alone, the locus of /d/ overlaps with that of /g/ but they become distinct when F_3 is also taken into account.
- The vocalic gestures are obtained by the four-region DRM with the first two formants inside the range where the area and formant frequencies are linearly related (between about 0.5 and 10 cm^2, the OTM mode), whereas the plosive consonant gestures are obtained by the eight-region DRM model with the first three formants also including the area range between 0 and 0.5 cm^2 (i.e., the closing-opening area, see Chapter 4, the TM and TTM modes), a range where the relation between area and formant frequency is non-linear (Mrayati et al., 1988; Carré and Chennoukh, 1995).
- One question remains: given that the consonant places of articulation are obtained with the closed-open eight-region model, what happens to the places of articulation for [uCu] when using a closed-closed model? The behavior of regions under the closed-open and the closed-closed models is not the same (see Figure 5.2). However, in the production of natural speech, the places of articulation (i.e., the regions) remain stable across all degrees of lip opening.

Unfortunately, we know of no studies on the acoustic contrast between open and (quasi-) closed lips during the production of [u], and therefore the model, as is, could not deal with this particular case. Obviously, this point needs further investigation.

6 Vowel systems as predicted by the model

This Chapter shows that vocalic trajectories obtained by the model predict actual vowel systems observed in a large number of languages. The correspondence between actual and predicted vowel systems can be attributed to the DRM that shapes the acoustic plane into a limited number of crisscrossing vocalic trajectories. The actual vowels in the inventory of a given language always fall on these trajectories.

6.1 Problems of vowel system predictions

Languages of the world have been objects of numerous comparative studies to pinpoint the major language families and their characteristics, in an attempt to understand their evolution and the ways diachronic changes have occurred. One of the first objectives is to study the *phonological system* and, in particular, the vocalic system of any given language. How many and which vowels does the language contain? Although vowel inventories do exist (e.g., Crothers, 1978; Maddieson, 1984), they represent only lists and are not intended to predict or explain the vowel system in any given language. For that task, a number of methods have been proposed (Liljencrants and Lindblom, 1972; Lass, 1984; Lindblom, 1986b; ten Bosch et al., 1986; ten Bosch, 1991; Schwartz et al., 1997b; de Boer, 2000; Diehl et al., 2003; Oudeyer, 2005; Moulin-Frier et al., 2015). The approaches cited represent attempts to explain vowel systems from the functional characteristics related to production and perception, rather than simply from their formal properties. Most authors first define the maximum acoustic space in the F_1-F_2 plane based on an articulatory model that gives rise to the vowel triangle, and then account for the distribution of vowels within this space using some criterion (e.g., maximum perceptual dispersion). Diehl et al. (2003) show that discrepancies between predicted and observed vowel systems are quite small and could be assigned, on the one hand, to the general difficulty of predicting central vowels and, on the other, to the large number of vowels predicted on the [iu] axis. At the same time, most of these studies propose a solution only for a single system with a given number of vowels and fail to consider other possible systems.

In our view, there are specific problems with these approaches, such as:

- Taking into account only the maximum acoustic space to predict vowel systems assigns the same status to all points inside a vowel triangle that could be reached with the same ease. Although Lindblom places vowels in a *continuous* space (Lindblom, 1986b, p. 36), he structures this space in terms of *discrete*

DOI 10.1515/9781501502019-007

and hierarchically organized units. In contrast, in our approach outlined in the previous chapter, vowels are reached by dynamic transitions from one to another, using strategies derived from articulatory and acoustic properties of the vocal tract. Also, in a continuous acoustic space, the vowels obtained cannot be characterized as labial or non-labial if they are characterized only according to their formant frequencies. Instead, our dynamic approach leads to distinctive trajectories that possess gestural labels.

- Prediction of an intermediate vowel to be situated at the midpoint between two defined vowels will be correct only if it is on the vowel trajectories that structure the acoustic space. For example, when the tongue gesture corresponding to the [ai] trajectory in the F_1-F_2 plane is co-produced with the lip gesture leading to [ay], [y] is not situated at the acoustic or perceptual midpoint of the [iu] trajectory — it is closer to [i] (Figure 6.1). In contrast, the maximum acoustic dispersion criterion within an unstructured vowel triangle predicts a new vowel between [i] and [u] which is always situated at the acoustic or perceptual midpoint of the [iu] trajectory.
- As shown in Section 5.3, two central vowels, like /ɯ/ and /ø/, may have almost identical F_1-F_2-patterns but quite different production characteristics: one is obtained by a central tongue constriction without labialization, the other by a front tongue constriction with labialization. Even if their static F-patterns are nearly identical, the formant trajectories in the F_1-F_2 plane, from other vowels to these central vowels, are different (Carré and Mrayati, 1995). These two vowels cannot be differentiated in a static approach.
- It is not possible to account for the hierarchy of vowel complexity using an acoustic or a perceptual approach alone, even if it can be assumed that the perceptual system is well adapted to perceive characteristics of the acoustic tube. On the basis of perception, the "complexity" of the systems is proportional only to the number of vowels. But, in the production domain, one or two gestures can be active, and the more gestures involved, the more complex the vowel production is (for example, the production of [ay] with two gestures is more complex than that of [ai] with one gesture, see Lindblom, 1990a). In a static approach, based on the acoustic space alone, these complexities cannot be accounted for.

As a consequence of the present approach, the F_1-F_2 plane is structured and dynamically "discretized" by the acoustic properties of the tube. Should a system acquire more vowels, it would need more gestures, thereby increasing its complexity. But an efficient system would require that a new gesture be fully used. A new gesture is not assigned to a single new vowel; for example, the tongue gesture corresponding to the trajectory [ai] with the vowels [a, ɛ, e, i], co-produced with the lip gesture leads to *all* the vowels ([a, œ, ø, y]) on the [ay] trajectory, and not only to one [y]. Maximum utilization of gestures corresponds to a "maximum utilization of the available distinctive features," as stated by Ohala (1979, p. 185), and as observed by Clements (2003) in several languages. The use of "speech gestures" as deformations of the area function in production instead of "features" removes a complaint voiced by Lindblom (1986b, p. 41): "A major difficulty... is to give a substantive, *deductive account* of the features."

6.2 Prediction of vowel systems from DRM trajectories

Prediction of vowel systems belongs in the domain of phonology. Diachronic phonology tries to explain evolution of vowel systems. For example, Ohala (1981) proposes that new vowels arise from talkers mispronouncing and listeners "normalizing" or "correcting" existing vowels; in contrast, Lindblom et al. (1995) postulate an adaptive process. In our book vowel systems are predicted by the DRM, even if it will not predict all systems that exist in the multitude of real-world languages. Our objective is to push the deductive approach based on dynamic properties of an acoustic tube to its limits. As explained in Chapter 5, the model leads to intrinsic primitive vocalic trajectories (Carré, 2009a) that represent the framework on which vowel systems can be built, starting at the simple and gradually expanding to the complex ones (Figure 6.1). Thus, the primitive trajectories constitute the basic elements of the vowel coding system: the coding of two successive vowels is less costly for vowels pertaining to the same primitive trajectory than pertaining to two different trajectories. To improve coding efficiency, maximal use of a primitive trajectory is given preference (consistent with the "maximum utilization of the available distinctive features," as stated by Ohala, 1979, p. 185). A behavior like this is observed in Table 5.1 on the maximal use of the [ai] trajectory.

In agreement with the deductive logic we have adopted, vowels are selected according to three successive hierarchically ordered criteria: maximum acoustic dispersion, maximum use of each trajectory (i.e., maximum number of vowels on any one trajectory), and low complexity (i.e., low number of trajectories). The use of these criteria necessitates making decisions between possible choices: maximum acoustic contrast and an increasing number of vowels can lead either (1) to an increase in the number of trajectories or (2) to an increase in the number of vowels on a trajectory. The two choices are not mutually exclusive because, once a new trajectory is chosen, it will be generally used to its maximum, i.e., the number of vowels on the trajectory will increase. Although deciding between the two choices is not an activity expected to be performed by a deductive model (i.e., in

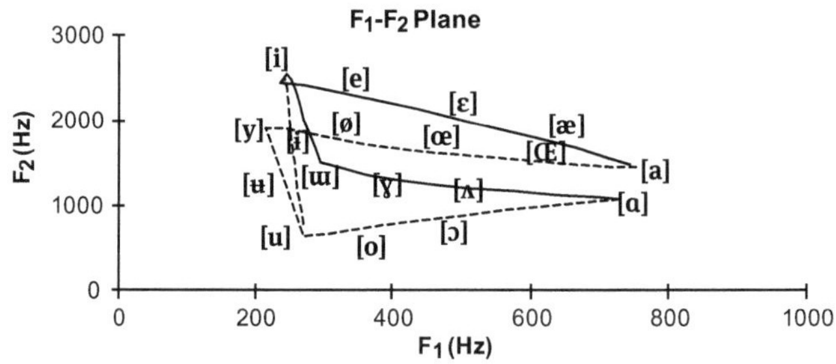

Fig. 6.1: Formant trajectories in the F_1-F_2 plane obtained by the DRM model used for predicting vowel systems.

order to expand the number of vowels in a system, one cannot predict whether the model will prefer to increase the number of vowels on one trajectory or add a new trajectory), still, allowing decision alternatives is not incompatible with a deductive approach. On the contrary, when the choice is to add a new trajectory, the availability of alternatives is able to explain the diversity of vowel systems – trees growing branches. From the tree of a given system having n vowels new systems could grow with n+1 vowels following either one of two existing branches more or less equivalent as regards acoustic dispersion and complexity. In fact, the choices arise as a direct result of the deductive approach. Scenarios like this are consistent with classifications proposed in vowel inventories (Crothers, 1978; Schwartz et al., 1997a, 1997b). For example, the vowel inventory of Crothers (1978) has two branches, one with and one without /ɨ/ (Figure 6.2). The most common vowel system consists of the three vowels /a, i, u/; for a four-vowel system, either /ɛ/ or /ɨ/ is added to the three. Without the possibility of choice, a deductive approach would lead to a single solution – no tree and no branches, as proposed, e.g., by Lindblom (1986b).

The acoustic tube and the DRM are agnostic as far as languages are concerned and, therefore, all gestures and trajectories in the model are *acoustic phonetic* e.g., [1]-[2], at first. When adopted by a language, a trajectory will become *phonological*, e.g., /ai/. To predict vowel systems going from simple to complex, the [ai] trajectory appears as the best first choice: the corresponding /ai/ phonological gesture to produce this trajectory involves only one phonetic gesture (by the tongue). Next, for the second choice, the [au] trajectory should be the best because it offers maximum acoustic contrast with respect to [ai]; the corresponding /au/ phonological gesture involves two co-produced phonetic gestures (one by the tongue and one by the lip). Although [iy], involving only one phonetic gesture, could have been the second choice, it would not have been the best because the acoustic contrast in it is relatively small (i.e., F_2 changes very little). With only the first two selected trajectories the following systems can be obtained by successively adding vowels that fulfill the criterion of maximum acoustic contrast between a new vowel and all those already present along the already adopted trajectories: [a, i, u], [a, i, u, ɛ], [a, i, u, ɛ, ɔ], [a, i, u, ɛ, ɔ, e], and [a, i, u, ɛ, ɔ, e, o].

Among the above listed systems, those with three, five, or seven vowels are acoustically well balanced and consequently stable. Therefore, we can suppose that they are more frequently found. A vowel system is acoustically well balanced (Maddieson, 1984, p. 138) and stable when its trajectories are fully used. Considering one trajectory that could be able to produce several vowels but produces only one, it will either disappear (the cost of keeping it would be too high) or more vowels on the trajectory will appear (consistent with the principle of maximum use of a feature).

In a system going from three to four vowels, [ɛ] instead of [ɔ] is chosen first because, for identical vocal effort, the amplitude of [ɛ] is higher than that of [ɔ]. Expanding to the five-vowel system, the vowels [ɛ] and [ɔ] are placed in the acoustic middle of the [ai] and [au] trajectories. Similarly, for the seven-vowel system [e] and [o] are placed in the acoustic middle between [ɛ] and [i] on the [ai] trajectory and between [ɔ] and [u] on the [au] trajectory. Vowels close to [a] are not chosen because their relative acoustic proximity makes them difficult to discriminate.

Thus, instead of extending the number of vowels on the [ai] trajectory by choosing [ɛ], the trajectory [iu] will be added leading to a new class of vowel system with three trajectories ([ai], [au], and [iu]) and a central vowel [ɨ] located in the acoustic middle between [i] and [u], generating the following five vowel sys-

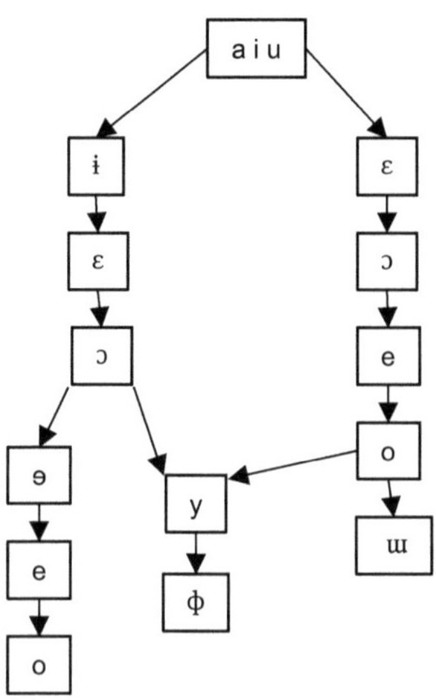

Fig. 6.2: Crothers' vowel inventory (1978, Figure 10, p. 115, with permission by Stanford University Press). The most common vowel system consists of the three vowels /a, i, and u/; then, for a four-vowel system, we observe an addition of the preceding vowels plus either /ɛ/ or /ɨ/.

tems: [a, i, u, ɨ], [a, i, u, ɨ, ɛ], [a, i, u, ɨ, ɛ, ɔ], [a, i, u, ɨ, ɛ, ɔ, e], and [a, i, u, ɨ, ɛ, ɔ, e, o]. In this case, it is the six-vowel system, but not the five-vowel system, that is acoustically well balanced.

Continuing the process of expanding the vowel systems, still guided by the criterion of maximum acoustic contrast, a new gesture will be necessary to increase the number of vowels, either by adding more vowels on the existing trajectories or by introducing a new trajectory. If a new trajectory is the option, the [ay] trajectory is obtained by having a labial gesture co-produced with the /ai/ gesture. The following three vowel systems are obtained: with the third using all the possibilities offered by the labial gesture: [a, i, u, ɛ, ɔ, e, o, y], [a, i, u, ɛ, ɔ, e, o, y, œ], and [a, i, u, ɛ, ɔ, e, o, y, œ, ø]. It is interesting to note that instead of [ay], the [aɯ] trajectory could have been chosen – representing another branch.

With the basic tongue and lip gestures, other complementary gestures, such as nasal/non-nasal, ATR/non-ATR (Advance Tongue Root, Ladefoged and Mad-

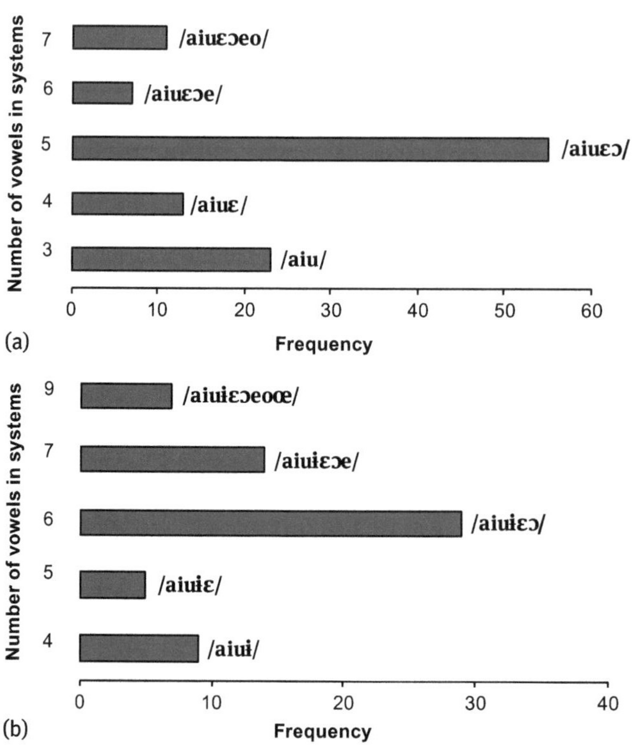

Fig. 6.3: Vowel systems as represented by the Crothers inventory (1978, Appendix III, pp. 138–143, with permission by Stanford University Press), (a) without [ɨ], (b) with [ɨ].

dieson, 1996), and others, can be added, leading to complementary sets of vowels. Sociolinguistic studies may explain why certain choices have been retained (Labov, 1972). The systems obtained according to our deductive approach can explain the classification of vowel systems according to the frequency of languages in the Crothers (1978) inventory – used by Lindblom to test his predictions (Lindblom, 1986b, p. 16) – and in the UPSID inventory used by Schwartz et al., (1997b, p. 273). In the Crothers inventory of 209 languages (Figure 6.3) we notice: (a) two main classes of vowel systems, one without [ɨ], and one with [ɨ] appearing after the three-vowel system /a, i, u/; (b) one system with $n + 1$ vowels obtained from the preceding one with n vowels by addition of a new vowel; (c) the acoustically well-balanced systems are the most common: the five-vowel system without [ɨ] and the six-vowel system with [ɨ]), (d) a trend of full use of the possible vowels on a trajectory. The same outcome is likely to be observed for using UPSID. With or without [ɨ], predictions based on the DRM are consistent with these results.

6.3 The phonology-phonetics relation

The good correspondence between observed and predicted vocalic systems permits discussing the relation between phonological symbol and the area function, i.e., the relation between phonology and phonetics. This relation is generally considered weak and mediated cognitively by some process of translation. For example, O'Shaughnessy (1996, p. 1729) noted the "lack of a simple relationship between many phonemes and their acoustic realizations." Nearey (1997) considered speech as weakly constrained by characteristics of the mechanisms of production and perception, and proposed a "double weak" concept. Concerning gestural invariance, Liberman and Mattingly (1985, p. 22) came forward with the assertion saying that "... the gestures have a virtue that the acoustic cues lack: instances of a particular gesture always have certain topological properties not shared by any other gesture." Lindblom (1996, p. 1689) questioned this assertion by stating that "The current evidence does not favor such a position... the prospect of finding articulatory invariance, or of showing that articulatory representations are richer and more distinctive than acoustic patterns, appears utterly remote." Incidentally, this statement can also be read as refuting Fowler's (1986) claim of "... both the phonetically structured vocal-tract activity and the linguistic information... are directly perceived (by hypothesis) by the extraction of invariant information from the acoustic signal... the signal is transparent to the phonetic segments." In fact, the present approach agrees with Fowler's thesis because it infers deformation gestures from acoustic properties inherent to the tube. These gestures lead to phonological entities, but they are also directly linked to the phonetic code. Actually, they are symbolic primitives that map into phonological systems because vowels and consonants are products of speech gestures. In fact, we find ourselves in the process of drawing the outlines of an *acoustic phonology* similar to the articulatory phonology that Browman and Goldstein (1992) derived from their data. Thus, as proposed by Fowler et al. (1980a), there is no mediating translation from symbol to articulation and acoustics. If the speech communication system is explained by physical properties of the acoustic tube that generate and modify formant frequencies, then this system supports the concept of a "single-strong theory" (everything deriving from the acoustic tube), instead of accepting the premises of a "double-weak theory" (Nearey, 1997).

6.4 Conclusions

The deductive approach based on a vocalic space with *structured* trajectories has enabled the *prediction* of vowel systems specifying the number of vowels they contain, as well as their stability, balance, and the frequency of their use. As consequences of this statement the following conclusions can be made:

Vowels generated by the combined acoustic tube-DRM engine confirm the overriding role of primitive trajectories obtained or defined by the model. Vowels on these trajectories are *dynamic byproducts* of the trajectories themselves, rather than static targets. Recall also that the main diphthongs observed in inventories (see, for example, Maddieson, 1984) are on the primitive trajectories.

The approach taken enables us to predict and explain not only vowel systems but, as shown in Chapter 5, also a system of consonants. This actually amounts to a positive resolution of a wishful comment on deductively predicting vowel systems by Ohala (1979, p. 185): "...It would be most satisfying if we could apply the same principles to predict the arrangement of consonants, i.e., to posit an acoustic-auditory space and show how the consonants position themselves so as to maximize the inter-consonantal distance...." In fact, the loci of the regions represented by the closed-open DRM operating in the F_1-F_2-F_3 space correspond to the standard places of articulation for plosive consonants (Mrayati et al., 1988; Al Dakkak et al., 1994; Carré and Chennoukh, 1995; Carré and Mody, 1997). The model accommodates a theoretical framework equally valid for vowel and plosive consonant production (Clements, 1992). The vowel systems predicted in Section 6.2 explain the generally accepted classification of data collected from natural languages. The good match between deductive prediction and observation makes it possible to formulate a process of mapping phonetic data to phonological data.

7 Speech dynamics and the model

This chapter focuses on the inherently dynamic nature of speech showing that information emerges from transitions rather than from successive static states. This dynamic representation confers an overriding importance to the dimension of time, movement, synchrony between primitive gestures, as illustrated in examples of vowel transitions and syllabic co-production. Experiments using pseudo-vocalic trajectories outside the vowel triangle serve as demonstrations that the slope and the duration of the transition are parameters sufficient for the perceptual system to identify, categorize, and discriminate vowels.

Through the previous chapters, we arrived at presenting the speech production process – gestures and systems of vowels and consonants– as dynamic phenomena. In the present chapter, this dynamic process will be ported into the time domain to specify issues related to the direction, range, and rate of formant transitions, as well as to gestural synchrony, kinetics, and the reduction of gestures. In Chapters 3 and 4, we used an acoustic tube and a model to map deformation gestures onto distinctive acoustic trajectories in the F_1-F_2 plane. Subsequently, we showed that this model is fully representative of the production of oral vowels and of the places of articulation of plosive consonants. Production of the syllable results from the co-production of vocalic and consonantal gestures (Chapter 5). As regards vowel production, vowel trajectories are characterized first by their direction. Figure 7.1 a shows the [ai] trajectory in the F_1-F_2 plane with its specific direction, an example from the set of primitive vocalic trajectories illustrated in Chapter 5. The question that arises is how to produce each of all possible V_1V_2 transitions to vowels lying on this trajectory. It is here that the predominant role of timing in our dynamic approach becomes clear. Let's take, [ai], the trajectory with the two intermediate vowels, [e] and [ɛ] on it. When producing the vocalic transitions [ai], [ae], and [aɛ] all with a 90-ms duration, they can be distinguished based on their different velocities (the dF_i/dt rate) along the [ai] trajectory in the F_1-F_2 plane. If the second vowel is recognized from the transition rate, then these rates are necessarily *distinctive* as regards the transitions (Figure 7.1 b). All along the transition, the rates carry information on the vowel in the process of being produced. However, in *real* production, the rate of transition changes over its duration, starting and finishing at a lower rate than the maximum reached in the middle (acceleration, steady state and deceleration). Figure 7.2 shows time derivatives of the first formant transition for [ai], [ae], and [aɛ] produced by the same speaker. For these three tokens the duration of the transitions is approximately constant (~180 ms), making it possible for a listener to identify the vowel targeted by the transition, provided he/she has acquired knowledge of the function repre-

DOI 10.1515/9781501502019-008

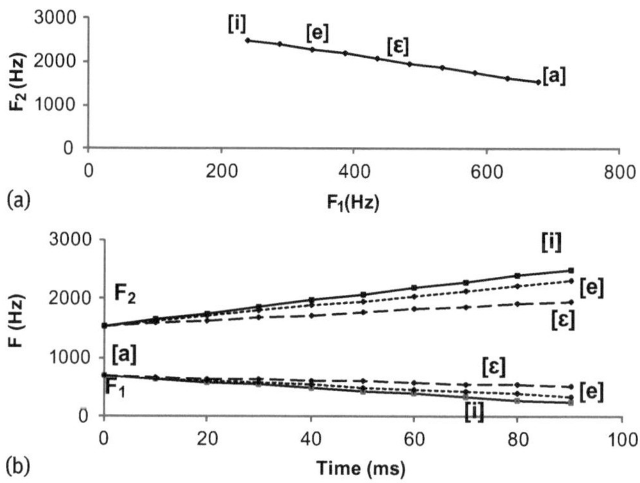

Fig. 7.1: (a) The [ai] trajectory in the F_1-F_2 plane. (b) F_1 and F_2 formant frequencies of the [ai] transition as a function of time, assuming a 90-ms transition duration.

senting rate change. Chapter 8 will show that this could indeed be the case. But, for predicting the vowel during the transition, the total transition duration must also be known.

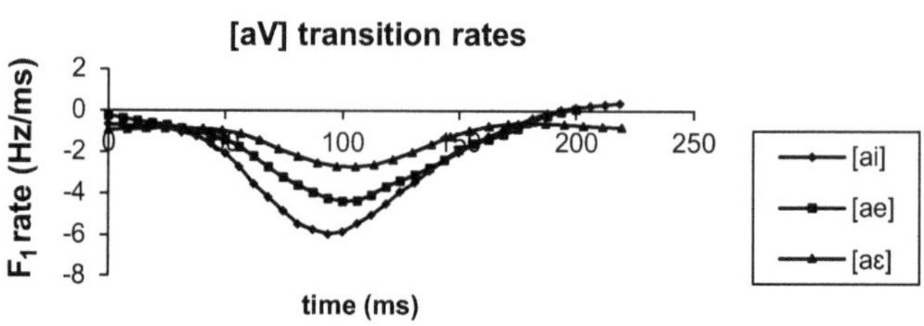

Fig. 7.2: F_1 transition rates for [ai], [ae], and [aɛ] produced by a single speaker.

Therefore, timing is a crucial parameter for defining a specific transition on a trajectory and the dynamic approach implies that we *know* this parameter. But what could be the *temporal unit* for the production of deformation gestures? First of all, the jaw makes a movement every time we pronounce a syllable. Lindblom et al. (1999) (see Figure 7.3) showed that the frequency of jaw movement requiring a minimum of input energy is between 3 and 7 Hertz (or between 140 and 300 ms, the duration between a short and a long syllable). The variability of the precise duration derives from individual speaker characteristics, as the resonant frequency

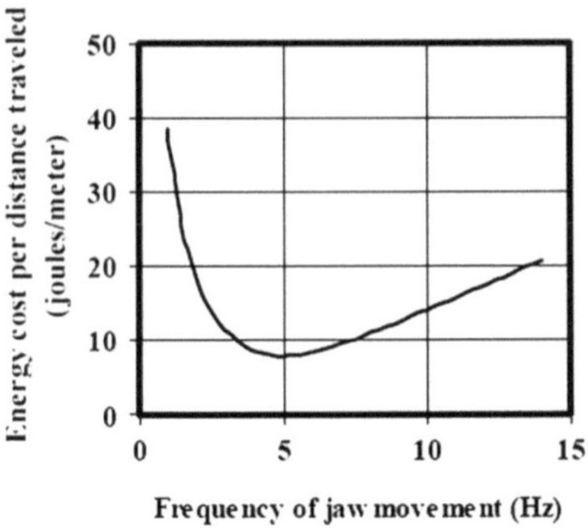

Fig. 7.3: Energy cost for jaw movement (from Lindblom et al., 1999, Figure 1, p. 403, with permission Charles University (Prague), Karolinum Press).

of the jaw depends on its weight, its volume, its muscles, etc., all of which vary from one individual to the next. To arrive at a reasonable estimate consistent with our minimal energy criterion, we assume that the duration corresponding to a single oscillation of the jaw is around 200 ms on the average – a figure that we shall regard as the mean duration of the syllable, in line with generally accepted norms (Greenberg, 1999). This jaw oscillation could have been used by humans according to some (e.g., MacNeilage's 1998 "frame") although doubted by others (e.g., Ohala, 2008).

In the same vein, the duration of the transition will permit defining the targeted vowel on the trajectory. Again, this duration will be speaker- and speaking style-dependent: fast speech will shorten and slow speech lengthen the temporal unit of all gestural deformations. After having introduced the timing parameters of formant transitions – its duration and rate of change – we will next discuss these concepts in more detail, in order to further clarify these dynamic aspects.

Targeted vowels are characterized and identified by their dynamic transitional movements of gestural and acoustic parameters and not by static targets: What is the link between gestural deformations and a specific formant transition? In a static vision we have two static vowels that define a transition, one at either end. The problem with this view is that the second vowel could be identified only at the end of the transition, which is not the case in the real world – vowel reduction and silent center experiments (e.g., Jenkins et al., 1994, as well as the identifiability of vowels under poor signal-to-noise ratio conditions presented in the next chapter) suggest that the final vowel is perceived well before it is reached. In contrast, according to a dynamic view the transition is defined by the movement pattern across the distinctive regions in the articulatory space and by the corresponding trajectory, that is, by the rate of deformation or formant frequency change over time. Thus, when dynamics of the transition are defined (i.e., when starting formant frequencies, the trajectory, the transition rate, and the syllabic rhythm are known), no prototypical static targets are necessary to identify the vowel at the end of the transition.

Such characterization of gestures may represent invariant elements in vowel reduction (Lindblom, 1963), consonant reduction (Duez, 1995), and hyper- or hypo-speech (Lindblom, 1990b). In other words, dynamic characterization can provide an explanation for various adaptation phenomena in situations where a sufficient acoustic contrast exists. It is well known that transitions are essential to characterize plosive consonants (Delattre et al., 1955; Dorman et al., 1977; Kewley-Port, 1982). It is also an accepted fact that transitions carry information in the case of vowel perception (Lindblom and Studdert-Kennedy, 1967; Strange, Verbrugge, Shankweiller, et al., 1976; Verbrugge and Rakerd, 1980; Nearey and Assmann, 1986; Di Benedetto, 1989b, 1989a; Nearey, 1989; Strange, 1989b, 1989a). Also, experiments have shown the importance of initial and final segments of transitions in CVC syllables for vowel identification in silent center experiments (Strange et al., 1983). These results contribute valuable information to the debate on the perception of vowel transitions, that is on which of three hypotheses to accept – the dual initial target plus final target hypothesis, the initial target plus slope hypothesis, or the initial target plus direction hypothesis (Nearey and Assmann, 1986; Pols and van Son, 1993).

Our own deducted hypothesis is that once the acoustic starting point in the F_1-F_2 plane as well as the direction and the rate of the formant transition are known, these parameters are sufficient for identifying the targeted vowel. Consider, for example, the production of [ae], starting from [a]. At the very beginning of the transition, the direction of the trajectory suggests that the targeted vowel is situated on the [ai] trajectory and the rate of formant frequency change depends on whether the targeted vowel is [ε], [e], or [i]. Since we know what the duration of the transition is – Kent and Moll (1969), Gay (1978), and Flege (1988) state that it is constant – the vowel may be identified *already at the onset* of the transition and also *anywhere along its course*. While this approach is able to accommodate speaker differences, vowel reduction, and noisy environments, so far it has not been sufficiently tested in production or perception experiments.

Figure 7.4 shows maxima of [aV] transition rates presented in the F_1-F_2 rate plane for ten normal and ten fast utterances produced by a single talker (Carré et al., 2007), with [V] being one of the French oral vowels. The variability may be due to the fact that the duration of the utterances was not controlled and that, to investigate transition velocity, temporal normalization of the utterances may be necessary.

Data and the reflections presented so far in this section lead to the following preliminary conclusions:

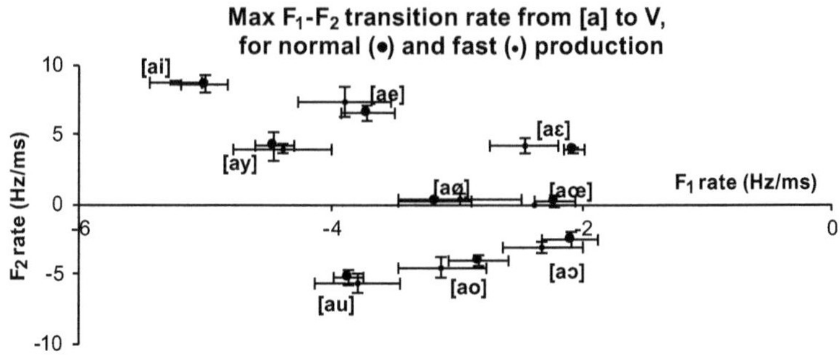

Fig. 7.4: Transition rate maxima (means and standard deviations) for [aV] produced ten times by speaker EM (normal (N) and fast (F) production). Note that [ai] has the highest velocity for both F_1 and F_2 on its trajectory, followed by [au] and [ae].

- The vowel production process is a dynamic movement characterized by formant transition direction, duration, velocity, and acceleration.
- The *temporal unit* of production is the syllable obtained by co-production of gestures.
- *Transition durations* are practically constant within a speaker- and speaking rate.
- *Transition rates* are essential characteristics of speech sounds.

These four conclusions can be extended to the perception of consonants, since it has been shown (Delattre et al., 1955) and widely accepted that formant transitions are crucial for the recognition of consonants.

Furthermore, we consider the deformation gestures modeled by the DRM as primitives of a communication system. If they are indeed primitives of the production and perception of speech, then these gestures should be independent from one another and may even be autonomous. The above preliminary conclusions were tested and confirmed in a series of perceptual experiments on vowel-to-vowel and VCV sequences (Carré, 1999, 2008, 2009b), but these experiments using French stimuli should be followed up. In the following sections we will present some preliminary experiments conducted with the aim of testing our predictions. However, to fully test these predictions, an experiment involving a larger subject pool would be required.

7.1 Characteristics of speech gestures as dynamic phonological primitives

In Chapters 4 and 5 we argued and demonstrated that dynamic gestures represent primitives – elements of the code for speech production. The question is whether they are also primitives of speech perception, that is, whether gestures are perceived as such and whether they, in fact, embody the code of speech communication, as several authors have suggested (Liberman and Mattingly, 1985; Fowler and Rosenblum, 1989; Mattingly, 1990). Should this be the case, the gesture would be perceived without mediating translation and a high degree of perceptual invariance is expected to overcome the relatively large variability of gesture characteristics – gesture duration, gestural movement (kinetics), and inter-gestural timing. While such invariance is consistent with the view that the perceptual representation of utterances is gestural, it provides only a necessary, but not a sufficient, proof for the fully fledged gesture perception theory proposed by Fowler et al. (2016). Nevertheless, it should be noted that the large range of variability may be the default condition when two speakers communicate as well as when an infant learns a phonetic repertory, and it may also be an indispensable ingredient for the process of phonological sound changes. Therefore, they deserve to be called "speech gestures."

7.1.1 The issue of duration range

When the DRM produces an [ai] token, the formant trajectory crosses the regions of the French vowels [ɛ] and [e] (see Figure 7.5 a), even if the presence of these intermediate vowels is not perceived in a natural utterance of an [ai] token. Percepts of four French listeners elicited by [ai] tokens with gesture durations ranging from 50 to 300 ms in steps of 50 ms is shown in Figure 7.5 b. The duration of the first vowel was 100 ms and that of the second 150 ms. The vowel complex /ai/ was perceived with gesture durations ranging from 50 to about 180 ms, whereas for longer durations an /aɛi/ percept was reported. In other words, at durations longer than 180 ms an additional intermediate vowel, /ɛ/, was heard despite the absence of a segmental marker. The 180-ms duration is within the syllabic range and, if phonetic memory also were to have such a span, it could account for the results.

Conversely, the same result is also fully consistent with perceiving a change in the *transition rate*: the rate is slower for [aɛ] than for [ai], as shown in the preceding subsection. A vowel encountered when the transition reaches the end of the time window that the 180-ms phonetic memory span represents could then automatically trigger a label even without the presence of a segmental marker. Thus,

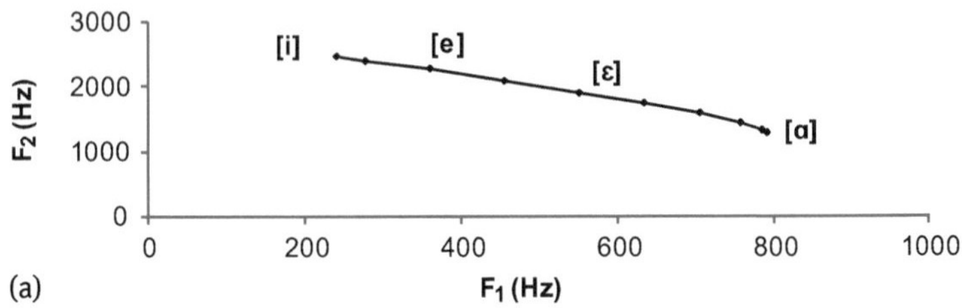

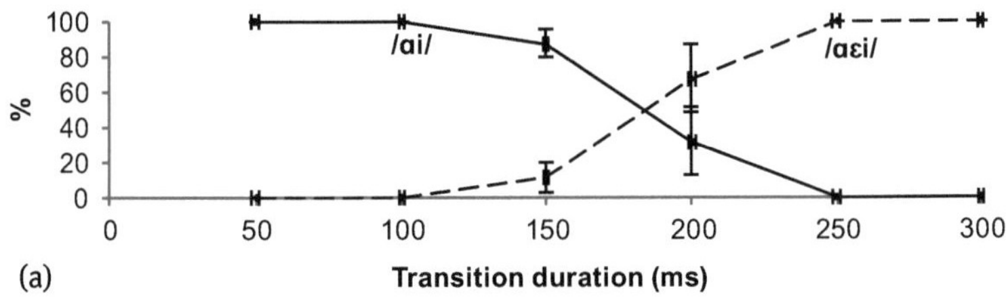

Fig. 7.5: (a) [ɑi] formant trajectory; (b) mean percent and standard deviation of /ɑi/ (solid lines) /ɑɛi/ (broken lines) as a function of gestural duration (from 50 to 300 ms in steps of 50 ms) for five French listeners. (From Carré, 1999, Figure 7, p. 644, with permission by the International Phonetic Association).

our syllabic co-production model seems to be adapted to a syllabic language that includes French as well. We can also interpret the results of this experiment by pointing out that /ai/ is perceived within a large range of transition durations: between 50 and about 180 ms in our example. The same observations also hold for transitions involving vowels other than [a] (Carré et al., 2001). Obviously, such results would not be possible without the listeners having, stored in their memory, a prototype of each of the possible vowels at the end of the transition. However, such phonetic memory appears to be listener-dependent: for a single listener the variability is smaller than the one for the combined data of several subjects (Figure 7.6).

In order to gain a better understanding of the perception of an intermediate vowel during V-V transition, a perception test with different transition durations was conducted on four native English speaker listeners (Carré et al., 2001). The 50% crossover for perceiving the intermediate vowel is obtained at a much longer transition duration (about 400 ms) than for native French speakers (about 180 ms). It is thus tempting to relate these results to those of Cutler et al. (1986, 1989, 1992) showing that French listeners perform rhythm-driven syllabic segmentation, while segmentation by English native speakers is controlled by stress. On the basis of these results, the possibility arises that segmentation of V-V sound streams is performed in accordance with the listener's phonetic experience of ex-

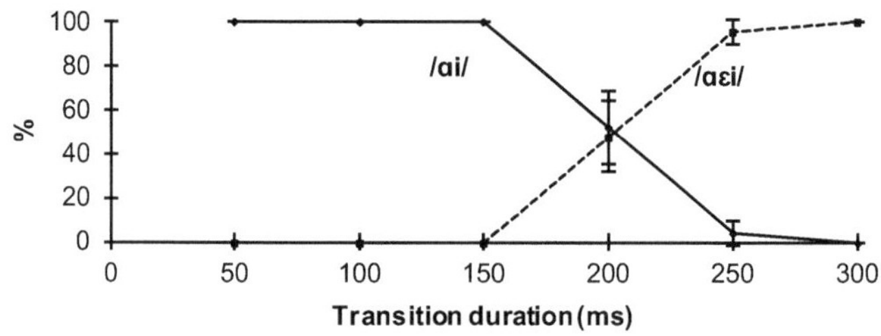

Fig. 7.6: Percentage (mean and standard deviation) of /ɑi/ as a function of transition duration for one French listener. The standard deviation values are generally smaller compared to the ones obtained for several listeners. (From Carré et al., 2001, Figure 6.5 b, p. 172, with permission by Karger Publishers).

tracting language-specific attributes from speech sounds (Johnson, 1997), that is, syllabic segmentation in French (about every 200 ms) and inter-stress segmentation in English (about every 400 ms corresponding to the average duration of two syllables). Segmentation thus relies on acoustic landmarks, such as rapid spectral change or rapid prosodic variation, i.e., changes in fundamental frequency or amplitude (Stevens, 1985). However, since in our experiments there is no specific acoustic landmark indicating a syllable- or inter-stress boundary, one could consider the possibility that the listener is guided by language-specific phonetic experience and detects the vowel corresponding to the parametric values of the signal at the expected boundary.

Would the results of this experiment be different if the listener had access to a syllable-duration referent on which the time scale would be normalized? Such a referent may be established when the listener hears a particular talker and adapts to the specific syllabic rhythm proper to his or her language, dialect, and idiosyncratic speaking style. Confirming this hypothesis, results by Miller and Liberman (1979) demonstrated that, when transition durations are manipulated, the passage of the perception from /ba/ to /wa/ will depend on syllabic duration.

In an attempt to modify syllabic rhythm, we repeated the former experiment with different durations of the [ai] transition using longer [a] and [i] vowels. The effect of lengthening the vowels from 100 to 150 ms was negligible. However, when we added an [ai] transition prime 300 ms before the stimulus proper, with the transition of the prime having a duration of either 50 or 150 ms, a shift in the boundary between the percept of two vs. three vowels was significant (Figure 7.7). These results suggest the occurrence of a *perceptual normalization* based on *transition rate*, rather than syllable duration. At the start of an encounter with a new speaker, or in experiments without syllabic rhythm, the listener bases his or her perception relying on a newly formed referent.

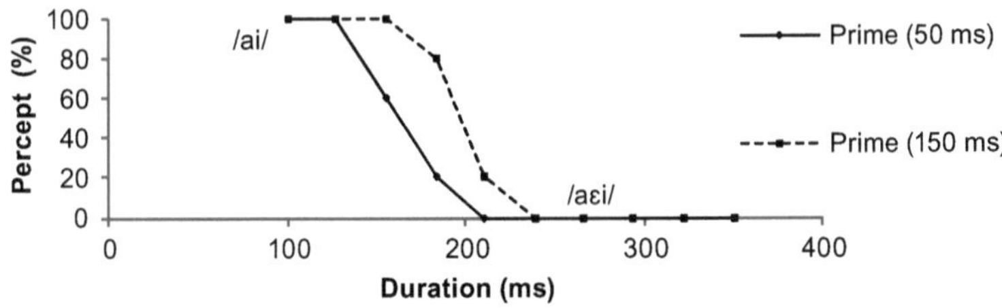

Fig. 7.7: Effects of a prime on the percept of an intermediate vowel on the [ai] transition.

At this point we felt compelled to replicate the experiment by Miller and Liberman on the perception of gradually lengthened V-V transitions, for which we used the extended temporal range investigated in the experiments shown in Figures 7.5 and 7.6. Results of this experiment for a single listener are illustrated in Figure 7.8. The percept for the [ua] trajectory (1) between 10 and 70 ms is /ba/; (2) between 70 and 120 ms is /wa/; (3) between 120 and 220 ms is /ua/; (4) between 220 and 350 ms is /uɔa/; and (5) beyond 350 ms it is unnatural but mostly resembles /uoɔa/. Since they demonstrate the importance of syllabic rhythm, these results are consistent with the findings of the "silent center" CVC experiments by Strange, indicating that identification of the vowel becomes more difficult when the duration of the silent vowel center is modified.

The major conclusion of this section is that large changes in transition duration do not affect the perception of primitive gestures. Furthermore, it allows us to propose the following:

- The range of transition durations is speaker- and language-dependent. Within a certain range, duration is essential for identifying the transition as being consonantal, diphthong, or vowel-to-vowel.
- Perception and identification of a transition having a given duration is a speaker–listener temporal adaptation process.
- Introducing transition prime leads to a change of the perceptual reference for transition rate (Verbrugge et al., 1976).
- These results are in agreement with well-known observations.

7.1.2 The issue of kinetics

Having gesture primitives implies that, within certain limits, the time course of speech gestures (temporal interpolation functions for gesture transitions) should be essentially irrelevant. This hypothesis was tested on the production of [ɑbi]

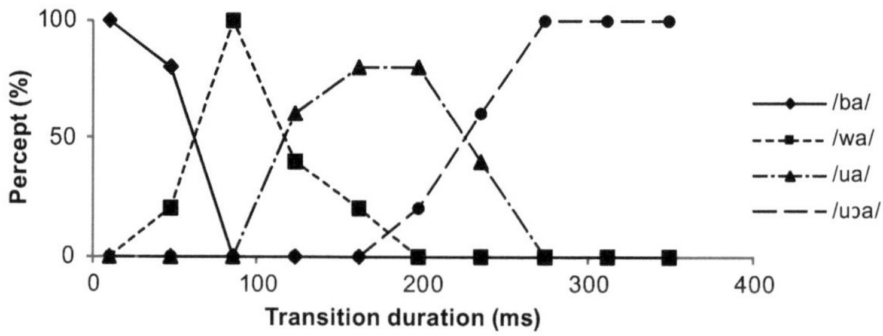

Fig. 7.8: Percentage of /ba/, /wa/, /ua/, and /uɔa/ percepts, as a function of transition duration.

synthesized with the DRM, that is, using commands to generate gestures (Carré and Chennoukh, 1995). Figure 7.9 shows formant behavior when using three different time interpolation functions for region-area gesture transitions to and away from the consonant (linear, logarithmic, and cosine interpolation). For these tokens tongue gesture and lip gesture were strictly synchronized. Although the different gesture transitions are reflected to some extent in the behavior of formant transitions, results of consonant identification tests were essentially identical for the three transition functions, despite the clearly visible divergence in the direction of the second formant transition at the end of the vowel [ɑ]. We should recall that during the [b] occlusion the token has essentially no acoustic energy, that is, the listener hears an abrupt termination of the transition from [ɑ] and an abrupt onset of the transition to [i].

7.1.3 The issue of gesture synchrony

To explore the perceptual effect of timing of simultaneous gestures, tokens with transitions generated by more than a single gesture are required. In French, the [ay] transition is obtained by the co-production of tongue and labial gestures, [abi] by two gestures (the vocalic and the b-consonant gesture), [aby] by three gestures (the tongue plus the labial gestures for the vowel and the b-consonant gesture), and [øby] by two gestures (the tongue and the [b] gesture, because both [ø] and [y] are labial vowels and thus the [øy] transition is produced by a single gesture). The following paragraph describes two experiments that examined the perceptual effect of gesture synchrony (Carré, 1999).

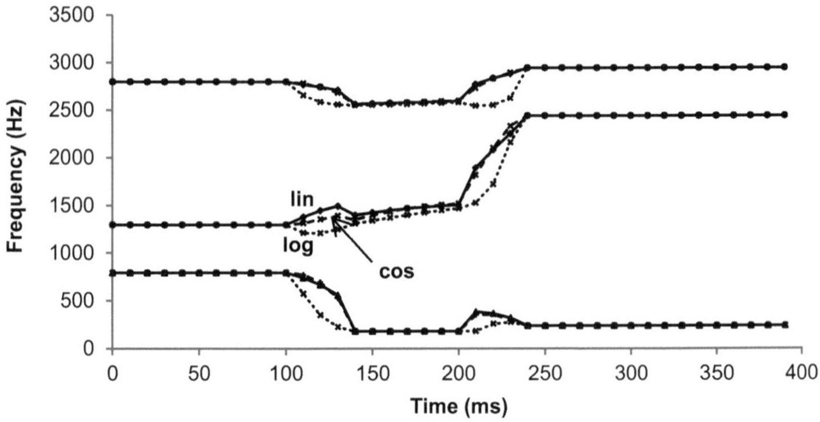

Fig. 7.9: Behavior of F_1, F_2, and F_3 in tokens of [ɑbi] synthetized using the DRM with three different gesture interpolation functions. (From Carré and Chennoukh, 1995, Figure 12, p. 239, with permission by Elsevier Publishing Co.).

[ay] tokens. As indicated above, [ay] is generated by the co-production of the tongue and the lip gestures. To investigate their perception, tokens were synthesized with the DRM such that while the duration of both gestures was fixed at 100 ms, the onset of the labial gesture with respect to the onset of the tongue gesture varied from a 60-ms lead to a 60-ms lag in 20-ms steps. Figure 7.10 a shows the timing diagram of the two gestures where negative asynchrony refers to anticipated and positive to delayed labialization. Formant transitions in the time domain were represented by a cosine interpolation of the gestural area. Figure 7.10 b shows formant trajectories for –60, 0, and +60-ms asynchronies of the labial gesture's onset. The –60-ms case represents a labial lead (anticipation), i.e., it is the labial gesture that is activated first (leading to a decrease of F_1, with F_2, known to be relatively insensitive to labial movements, remaining approximately stable), after which the tongue gesture becomes dominant, in order to reach [y]. The +60-ms case corresponds to a labial lag, i.e., it is the tongue gesture that is activated first but it is aimed at [i] before turning toward [y], due to the delay of the labial gesture. When the two gestures are exactly synchronous, i.e., in the 0-ms case, the formant trajectory is first aimed at [i] before swerving to aim at [y]. This means that, in fact, when the onset of the two gestures is simultaneous, the *acoustic effect* of the labial gesture is delayed with respect to that of the tongue gesture. Figure 7.10 c illustrates *perceptual results* showing that an /ay/ percept is reported for labial asynchrony between –50 and +10 ms, that is over an extensive range. Because the acoustic trace of the onset of the tongue gesture occurs before that of the labial gesture, the identification function is not centered at 0 but biased to-

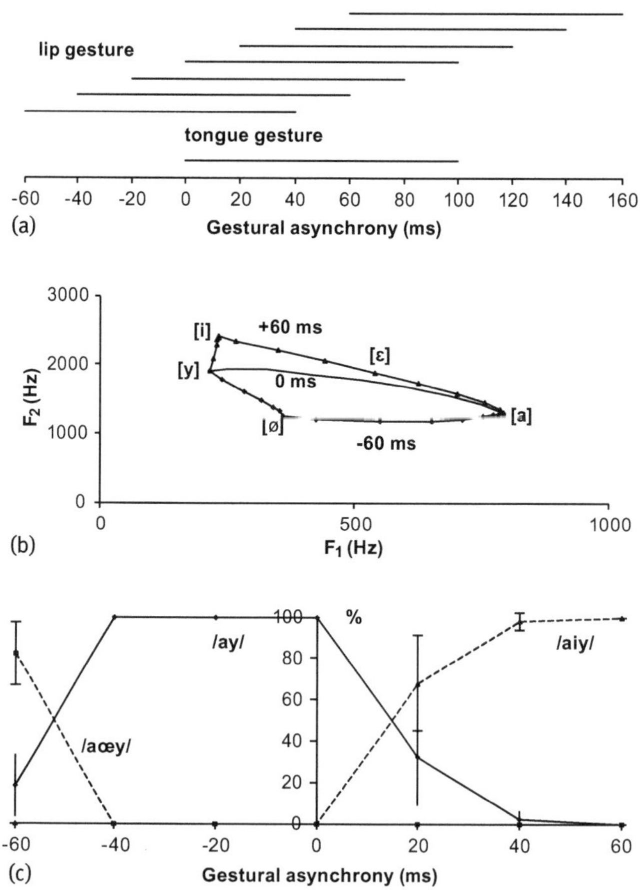

Fig. 7.10: Panel (a): Tongue and lip gesture asynchrony. Panel (b): Corresponding formant trajectories for three cases of labial phasing of −60 (square), 0 (diamond), +60 ms (triangle). Panel (c): Percent of /ay/ percept (and standard deviation) as a function of gestural asynchrony (from −60 to +60 ms in steps of 20 ms). (From Carré, 1999, Figure 8, p. 645, with permission by the International Phonetic Association).

ward labial anticipation. When the asynchrony is more negative than −50 ms, the percept is /aœy/ and when it is more positive than +10 ms, it is /aiy/. The main finding is that /ay/ is perceived over a large range of gestural asynchrony.

[abi] tokens. Figure 7.11 represents production of [abi], V_1CV_2 example. Figure 7.11 a illustrates the timing diagram while Figure 7.11 b shows the acoustic effect of the relative timing of vocalic and consonantal gestures and, consequently, of the synchrony/asynchrony of the resulting formant transitions. When the [ai]

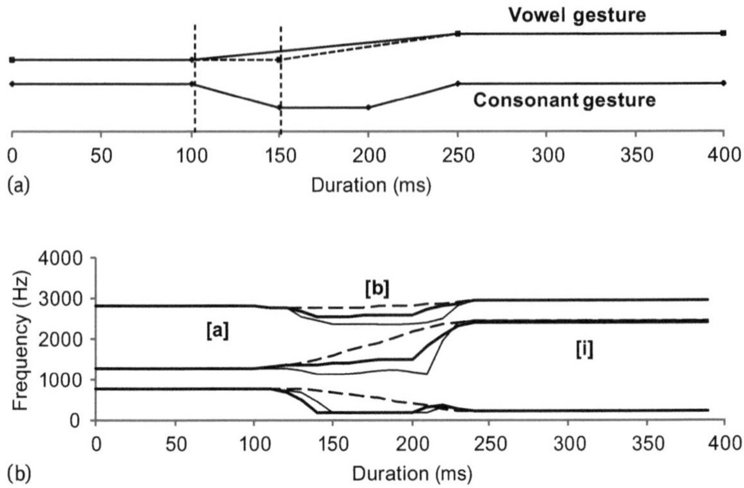

Fig. 7.11: Panel (a): Timing diagram of the consonant gesture superimposed on the vowel gesture with two relative timing schemes. Panel (b): Formant transitions for [ai] (dotted line). For [abi] (thick solid line), the consonant gesture is superimposed simultaneously with the vowel gesture. For [abi] (thin solid line) the beginning of the vowel gesture corresponds to the beginning of the full consonant closure.

transition and the [b] transition are simultaneous, the token [abi] is obtained: the C-V_2 transitions are influenced by the first vowel V_1 and, reciprocally, the V_1-C transitions are affected by the final vowel V_2, as Öhman has proposed (1966b). In contrast, when the onset of the vowel gesture toward V_2 is synchronous with the instant at which the consonant closure is complete, the transition V_1C in [abi] is not influenced by V_2 (thin solid line), just as Gay (1977) has observed.

A perception experiment for the case of co-production with two gestures was designed to examine the production of [abi], as illustrated in Figure 7.12 together with the results. The [b] consonant was chosen to avoid lingual coarticulation between vowel and consonant gestures. Durations of the consonantal closing, the full closure, and the opening were set to 50 ms each. The V_1CV_2 signals were produced without the plosive burst. The asynchrony of the two gestures was varied from –75 to +75 ms in 25-ms steps, where –75 and +75 designate a 75-ms anticipation and delay of the consonant onset, respectively, and 0 corresponds to exact simultaneity of the two gestures (Figure 7.12 a). The F_1-F_2 plot of three tokens, at –75, 0, and +75-ms asynchronies is illustrated in Figure 7.12 b. The curves for the two extreme asynchronies clearly show an abrupt change when /ɛ/ is reached, suggesting a break that may be audible. The perceptual effect of gesture asynchrony on perceptual boundaries is shown in Figure 7.12 c. The percept /abi/ was reported

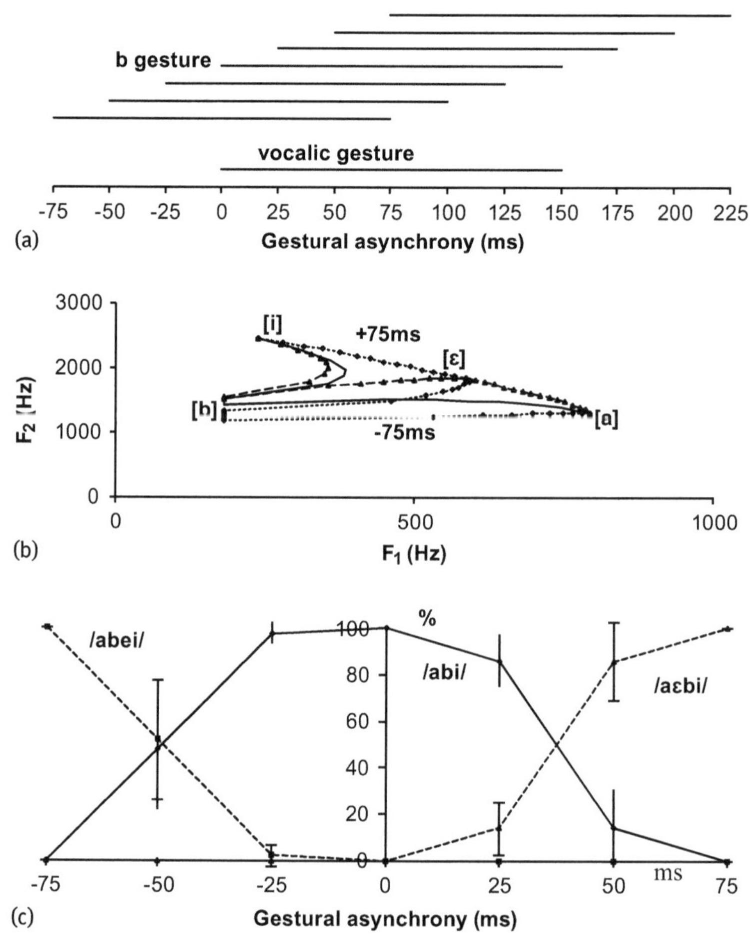

Fig. 7.12: Panel (a): Gestural durations, vocalic and consonant gesture asynchrony. Panel (b): Corresponding formant trajectories in case of asynchrony equal to −75 (diamond), 0 (bold), +75 ms (triangle). Panel (c): Percept (in % with standard deviation) of [abi] tokens as a function of gestural asynchrony (from −75 to +75 ms in steps of 25 ms). (From Carré, 1999, Figure 9, p. 646, with permission by the International Phonetic Association).

between [b] anticipations of approximately −50 ms and delays of +35 ms. A broad range of asynchrony is seen to lead to a constant [abi] percept, while listeners reported hearing /abei/ for asynchronies under −50 ms and /aɛbi/ for asynchronies above +35 ms, both percepts potentially due to the break in the otherwise monotonic transition. Thus, as a function of asynchrony, a new “phonetic vowel” ([e] or [ɛ]) can appear and could indicate a possible sound change process.

7.2 Formant transition rates

As we said earlier, the velocity of transitions is related to the individual duration of gestures. We have seen that whenever the transition durations are approximately constant (while still a function of the individual speaker and perhaps also of certain categories of sounds), the vowels V_1V_2 can be identified already at their onset, relying on the direction of the formant trajectory in the F_1-F_2 plane and on the transition rate (velocity) along that trajectory. Therefore, in this subsection we will investigate the role transition rate may play both in the perception of vowels and consonants and in vowel reduction (VR).

7.2.1 Vowel-to-vowel transition dynamics

In the following paragraphs we will discuss the direction of V_1V_2 trajectories, vowel and gesture reduction, and the perception of vowel transitions outside the vowel triangle. As will become apparent, the argument common to these three phenomena is the overarching importance of vowel transitions along distinctive trajectories. This view contrasts the generally still popular view of considering vowels as static elements of speech[1]. We hope that the following sections will be able to convince the reader of the contrary.

7.2.1.1 Reduction

Numerous classic studies exist on the properties of vowel formants (Chiba and Kajiyama, 1941; Peterson and Barney, 1952; House and Fairbanks, 1953; Fant, 1973), properties that have been seen to vary as a function of speaker, style, prosody, and phonetic context. These and other observations have led to the finding that vowels produced with sometimes markedly different formant characteristics could be perceptually equivalent. Because this multiple-valued nature of vowels was recognized early on, it has been generally assumed (first implicitly and, much later, explicitly, see e.g., Kuhl and Meltzoff, 1984) that the average formant values of vowels in isolation represent *targets* that the talker, during continuous speech, always strives, but often fails, to reach. Cases when vocalic targets are not reached have been termed articulatory *undershoot*, from the standpoint of production,

1 For a phenomenon as important as the production and perception of speech, it is difficult for us to accept that successive targets (= static positions of the vowels in the acoustic plane) would be only rarely reached in spontaneous speech. If this were the case, how would such a referential system find its place in the human production and perception system? In addition, to never reach such targets would be psychologically most troubling.

(Brownlee, 1996; Lindgren and Lindblom, 1996), perceptual *overshoot* from the standpoint of perception, and *vowel reduction* from the standpoint of acoustic trajectories (Lindblom, 1963). Vowel reduction is observed, for instance, in fast speech (where the vocal tract is not given enough time to complete the required task and, therefore, it is denied the opportunity of having its shape conform to that of an intended target) as well as in "hypo-speech" (Lindblom, 1990b). Then, the intended targets are recovered by the listener either by way of a special compensatory mechanism that produces a "perceptual overshoot" (Lindblom and Studdert-Kennedy, 1967) or by using previously learned and stored templates of vowel formant patterns that include the undershoot effects (Johnson, 1997). While such a mechanism would predict a one-to-one correspondence between production and perception of unreached target vowels, some results appear to cast a shadow of doubt over this prediction. For example, in production, vowel targets are often reached even in fast speech (Kuehn and Moll, 1976; Gay, 1978; van Son and Pols, 1992), while normal-rate speech does not necessarily ensure that the targets will be reached at all times (Nord, 1986). The situation is not less ambiguous in the perceptual domain: under certain conditions listeners seem to base their judgment on the *average* of the formant values covered by the trajectory. Since averaging the frequencies of a formant trajectory inevitably displaces the percept of the final frequency toward the initial frequency, such cases explain perceptual *undershoot* (van Son and Pols, 1993), rather than produce a perceptual compensation for vowel reduction. Other authors argue that first formant transitions are averaged and second formant transitions lead to a perceptual overshoot (Huang, 1987; Di Benedetto, 1989a). Investigations on aspects of vowel reduction related to both production and perception have identified numerous characteristics of the phenomenon in diverse phonetic contexts and at different speaking rates (van Son and Pols, 1993; van Wieringen and Pols, 1991).

Some investigations of V_1V_2 transitions with vowel reduction view the phenomenon from the perspective of a kinematic approach to speech that, in agreement with Strange (1989b), we also favor (Carré et al., 1994). The following subsection describes experiments on gesture reduction, on speech gestures as defined in Chapters 4 and 5 and used in the production and perception of vowel sequences, as well as on the production and perception of "vowels" simulated outside the vowel triangle. In all these experiments the DRM has been used to generate speech.

The [ai] reduction experiment. While vowel reduction in context can be understood as being the byproduct of a linguistic process – a process that strives to achieve perceptual invariance across phonetic variations resulting from variations of the context itself – existence of a parallel auditory process leading to an identical outcome cannot be altogether discounted. Indeed, there are indications that dynamic variations of the spectrum can lead to perceptual overshoot-like effects also for non-linguistic stimuli (Divenyi et al., 1995; Divenyi, 2009). Thus, studying, even within a limited linguistic framework, the effect of parametric changes of a transition on the perception of an unreached target could reveal hitherto hidden properties of the perceptual dynamics of both speech and hearing. The objective of the experiments described here was to investigate temporal aspects of vocalic transitions, namely, the relationship between their duration and their velocity.

In the experiments by Divenyi and Carré (1998), stimuli were synthesized vowels (synthesized using a formant synthesizer) with a 100-ms V_1 first portion followed by a transition on a linear trajectory in the F_1-F_2 plane toward target vowel V_2. Six American English V_1V_2-transition complexes were investigated: [ai], [ia], [au], [ua], [iu], and [ui]. Eight tokens were generated for each vowel complex. In the first experiment, the tokens had a *constant duration* but the duration of the transition *toward* V_2 was gradually increased (i.e., its velocity decreased), with the result that, as the transition became shorter and shorter, the final F_1-F_2 frequencies of the tokens got closer and closer to those of V_2 (see Figure 7.13 a). In another experiment, the transition portion of the tokens had a *constant velocity* but the duration of each token was gradually cut back, with the result that the final F_1-F_2 frequencies of the tokens got closer and closer to those of the target V_2 as the transition became longer and longer (see Figure 7.13 b). An almost perfect trade-off was observed between transition velocity and transition duration, but the relationship was non-linear (Divenyi and Carré, 1998).

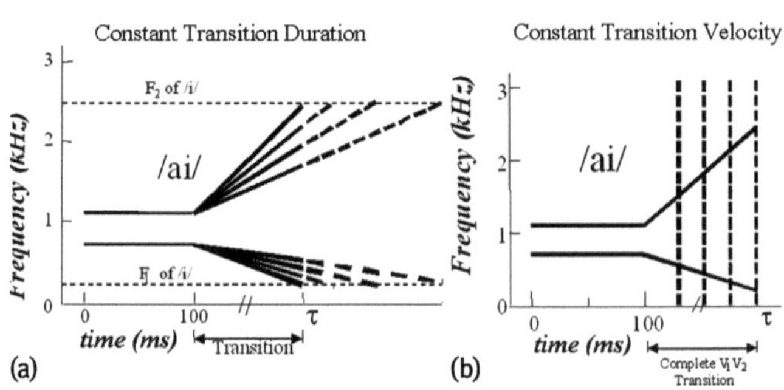

Fig. 7.13: Time frequency plots of the first two formants of the stimuli for the two experiments. (From Divenyi and Carré, 1998, Figure 1, p. 2957, with permission by the Acoustical Society of Amierica).

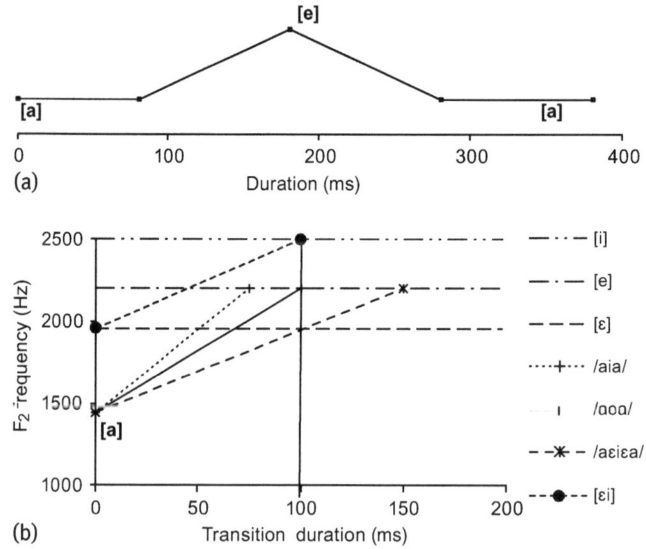

Fig. 7.14: Panel (a): Characteristics in the time domain of the [aea] succession to be tested (the transition duration is here 100 ms). Panel (b): Three different second formant transition rates from [a] to [e] and one transition from [ε] to [i]. The lower rate of the transition [ae] is essentially identical to that of [εi]. The transitions are perceived as /aia/, as /aea/ and as /aεiεa/ as a function of the rate going from fast to slow (after Carré, 2008).

Perceptual overshoot/undershoot for [aea]. If the rate of the transitions is used for the perception of a subsequent vowel, then it could account for the reduction phenomenon. To test this hypothesis, different $V_1V_2V_3$ items were obtained with a formant synthesizer with $V_1 = V_3 =$ [a], $V_2 =$ [e], V_1 duration = 80 ms, V_3 duration = 100 ms, V_1V_2 transition duration = V_2V_3 transition duration = 75, 100, 150 ms (see Figure 7.14 a displaying a transition duration of 100 ms after Carré, 2008).

When the duration of the transition was around 75 ms, /aia/ was perceived, that is an "overshoot" effect was obtained. When the duration of the transition was around 100 ms, the percept was /aea/, that is, realistic. But, when the transition duration was 150 ms, listeners reported hearing /aεiεa/. The question is what caused the vowel /i/ to be the percept in the conditions in which it was reported? Figure 7.14 b can help provide an explanation. Assuming that transitions are processed using a 100-ms integration window, the expected reference transition duration is 100 ms and any transition rate is evaluated in terms of what vowel the transition would reach at the end of the reference duration. A transition shorter than the reference, such as the 75-ms transition of [aea] in the figure, will be perceived with its slope prolonged to the vowel it would reach should its duration be

100 ms – [i] in the example. Therefore, with the 75-ms transition [aea] should be perceived as /aia/. Conversely, when the duration of the transition is longer than the reference, as the 150-ms transition in the figure is, the transition reaches /ɛ/ at the end of the reference duration and an /aɛ/ transition should be perceived (perceptual undershoot). Since the transition does not stop, the rate of that transition is established as the reference rate, and /ɛ/ becomes the new anchor. The continuing transition from this anchor is also assumed to have the reference duration of 100 and thus the transition points to /i/ at the end of the assumed 100-ms duration, resulting in a perceived /ɛi/ transition and an /i/ percept. While the overshoot effect (Lindblom and Studdert-Kennedy, 1967) could account for the perception of /aia/ at the short transition duration, it is unable to explain the /aɛiɛa/ percept at the long transition duration. In contrast, assigning a commanding role to the rate of transition can provide an explanation for both perceptual effects.

7.2.1.2 Perceptual integration of transition direction

The preceding experiment examined the reduction for the case of a vowel sequence generated by a single gesture, that is a tongue gesture [aea] corresponding to a rectilinear formant trajectory. But in the case of [aya] generated using the DRM two simultaneous gestures (i.e., tongue and lip gestures) are needed. If the two gestures are made strictly synchronous, (i.e., without labial anticipation), "lip acoustic effects" are obtained at the end of the lip gesture. The trajectory from [a] will point first to [i] before reaching [y] (Figure 7.15 a). In this case, reduction experiments on [aya] show that reaching [y] is not sufficient to perceive /aya/; in fact, the percept is /aia/. To perceive /aya/, it is necessary to stay on [y] for more than 20 ms (Figure 7.15 b).

This finding strongly suggests that vowel-to-vowel transitions are governed by temporal integration of the trajectory vectors in the F_1-F_2 plane. In fact, there appears to be a mechanism that computes the *time average* of the *length* and the *direction* of formant trajectories, which are by definition time-varying. Clear reduction effects can be obtained when the [aya] trajectory is rectilinear. Because, contrary to tongue gesture, the lip gesture will elicit a significant acoustic effect only when the opening is narrow (less than 1 cm^2), rectilinearity of the trajectory can be obtained only by an anticipated labial gesture.

7.2.1.3 Perceptual illusion: Vowels outside the triangle

If dynamics, in terms of direction and rate of formant transitions, is important for the perception of vowel-to-vowel transitions, then vowels should be perceived even on synthesized trajectories outside the vowel triangle as long as the direc-

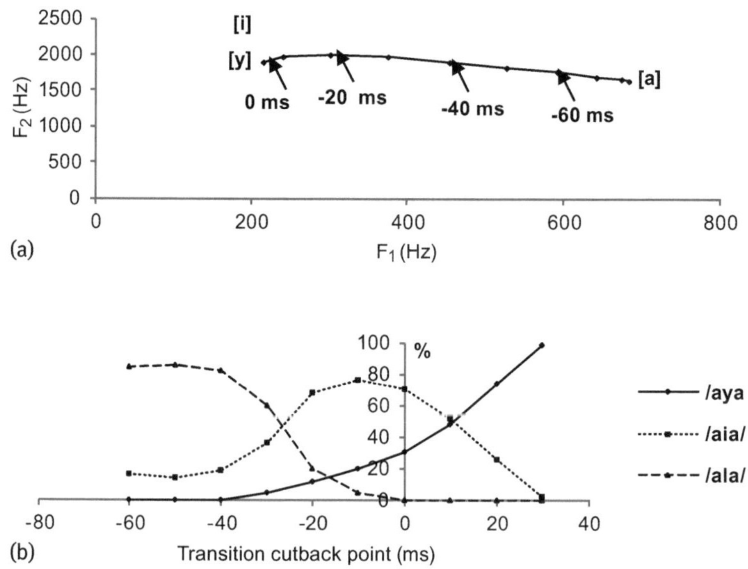

Fig. 7.15: Panel (a): F_1-F_2 plane representation of the [aya] transition for the eight tokens shown above. Note that the [aya] trajectory is curved. Panel (b): Results (average and standard deviation): Percent /aya/ (filled diamonds), /aia/ (squares), /ala/ (triangles) responses as a function of the duration of [y] or the transition cutback point (i.e., the point of return to [a]) where 0 ms refers to the condition in which the vowel [y] is reached but the transition immediately returns toward [a]. Note that 100% /y/ responses were obtained only for the 30-ms "positive cutback" condition, that is for the condition in which there was a 30-ms steady-state [y] before the transition actually turned back to [a].

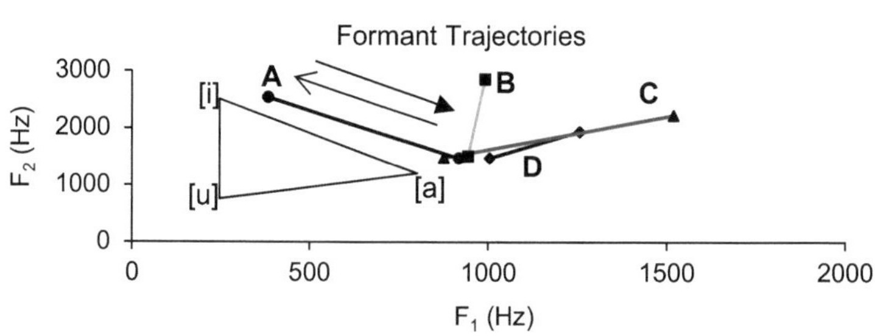

Fig. 7.16: The four trajectories (A, B, C, and D) in the F_1-F_2 plane. Note that the trajectories are always outside the vowel triangle. Their direction and extent was a function of the experimental condition.

tion and rate of transitions are the same as those for transitions inside the vowel triangle.

To test this hypothesis a test was conducted (Carré, 2009b). $V_1V_2V_1$ transitions A, B, C, and D were synthesized outside the vowel triangle. Perception of any of these transitions had to occur without reference to any known static vowel. The rates of all these formant transitions were similar to those of natural $V_1V_2V_1$ tokens (Carré, 2009b). The durations of the transitions were held constant at 100 ms.

Figure 7.16 shows in the F_1-F_2 plane the four $V_1V_2V_1$ formant trajectories A, B, C, and D outside the vowel triangle (Carré, 2009b). The A token that had the same direction and transition rate as an [ai] transition inside the triangle was perceived as /iai/ 71% of the time. Token B, which had the approximate direction and rate of [uɛ], was perceived 87% of the time as /ɛuɛ/. Tokens C and D that have the same direction and rate as [au] were perceived as /aua/ or /aoa/ 95% of the time. More precisely, token C alone was perceived as /aua/ 55% of the time, and token D, which had the same direction as [au] but a lower transition rate, was perceived as /aua/ 32% and as /aoa/ 63% of the time. The pseudo-vowel in the figure at which all four trajectories converge was perceived either as /a/, /u/, or /o/, depending on the dynamics of the transition. These findings show that our perceptual system assesses parameters of spectral change, specifically the rate and direction of frequency transitions (Divenyi, 2005a). To be able to accomplish this, it must be aware of the point of departure, but it is unclear whether this point is a phonetic or a phonological one, since neither of these disciplines deal with speech sounds outside the normal formant spectrum. Results of this last experiment suggest that the starting point of these trajectories may be understood using a phonological reference (Carré, 2009b).

Results of experiments on the perception of consonants outside the vowel triangle also indicate the existence of a similar illusion (Tran Thi et al., 2013).

7.2.2 Consonant-to-neutral vowel transition rates

Since the regions of the model are obtained from the rising and falling formant changes using a uniform tube as the point of departure, and since these regions correspond to the places of articulation and each region has a distinctive formant transition behavior (FTB), the following predictions can be made:

- To create phonetic distinctions from the uniform tube (which would produce the neutral vowel [ə], the "schwa"), the detection of the upward or downward direction of a transition is distinctive and sufficient.

- The rise or fall of the second formant can be used to separate /b/ from /d/ and /g/ also in transitions to and from the neutral vowel, thereby extending classic results by Liberman et al. (1957).
- The rising or falling variation of the third formant is used to separate /d/ and /g/.

Depending on the formant, a local deformation of the tube at the boundary between two regions may have no effect; for example, at the boundary between R_5 and R_6, a deformation has no effect on F_3, as extensively explained in Chapter 4. Several experiments have shown that the third formant is essential for differentiating consonants like /d/ and /g/. For example, Harris et al. (1958) note that the consonant /d/ cannot be obtained in an /i/ vocalic context without the third formant. Similarly, in categorical perception experiments, Godfrey et al. (1981) examined the boundary between /d/ and /g/ by changing the rate and direction of only the third formant transition, while the first two formants were sufficient for a /b/–/d/ contrast. In duplex perception, Mann and Liberman (1983) also used F_3 for the /d/, /g/ distinction. Lindblom (1996) displayed the F_2 and F_3 onsets of the plosives as a function of the second formant of the second vowel V_2 from Öhman's data (1966a). Indeed, there is a boundary between /b/ and /d, g/ in the plane of F_2 of the consonant onset as a function of F_2 of the vowel, and another boundary between /g/ and /b, d/ in the plane of F_3 of the consonant onset – F_2 of the vowel. Given the agreement between our theoretical predictions and these results, we decided to conduct an extensive study on the relationship between formants, vowel context, and the perception of place of articulation in French stops.

One of the listening experiments (Carré et al., 2002) examined the role of the second and third formant transitions on the boundaries between /b, d, g/ categories. Neutral vowel–consonant–neutral vowel tokens were synthesized following the scheme illustrated in Figure 7.17 a. The neutral vowel had vowel frequencies of 500, 1,500, and 2,500 Hz for F_1, F_2, and F_3, respectively. The experiment investigated the percept of the consonant when the excursion extremes of the F_2 and F_3 transitions were systematically varied by −1.5, −0.75, 0, 0.75, or 1.5 Barks with respect to the neutral baseline formant frequencies. At the extreme points of the transitions the frequency of F_1 was always 3 Barks (about $\frac{3}{4}$ of an octave) lower than that of the neutral baseline value. As the figure shows, formant transitions were linear and the consonant onset and offset portions always symmetrical in time. Four native French-speaking listeners were tested with 25 different tokens. Using a forced choice task, they had to indicate whether the consonant they heard was /b/, /d/, or /g/.

The reader should be reminded (see Chapter 5) that region R_5 corresponds to velar, R_6 to alveolar, and R_8 to labial constrictions, that is to the plosive consonants [g], [d], and [b]. Results of the experiment are illustrated in Figure 7.17 b showing, in the F_2-F_3 plane, the proportion of the positive (70% or more) and negative (less than 70%) percept of each of the three consonants. The 25 items are represented by different symbols, each representing a dominant labeling response. The results suggest that a decrease of F_2 and F_3 (corresponding to closing R_8) leads to a /b/ percept (square), an increase of F_2 and F_3 (closing R_6) leads to a /d/ percept (diamond), and a simultaneous increase of F_2 and decrease of F_3 (closing R_5) lead to a /g/ percept (triangle). Formants of the baseline neutral vowel F_2 = 1,500 Hz corresponds to the boundary between R_6 and R_7 while

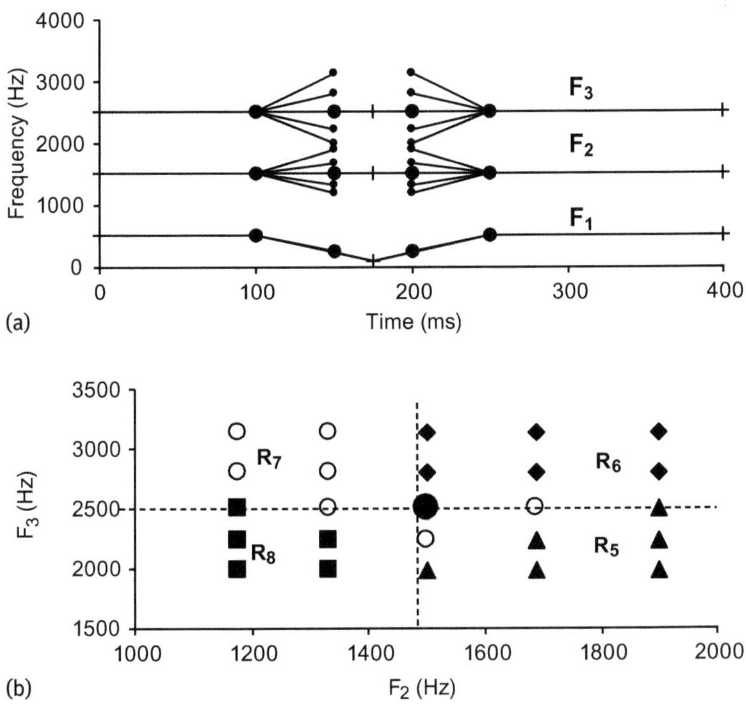

Fig. 7.17: Panel (a): Time-domain diagram of formant transitions of the stimulus for the perception of the consonant C in /əCə/ tokens (Carré et al., 2002, Figures 2 and 3, p. 151, with permission by the International Phonetic Association). Panel (b): /əCə/ results in the F_2-F_3 plane. The large circle represents F_2 and F_3 for the first and second vowel. Filled symbols correspond to stimuli for which there were responses at least 70% in favor of one category. Small circles correspond to less than 70%. The horizontal and vertical lines correspond respectively to flat F_3 and F_2 transitions. Decreasing F_2 and F_3 (corresponding to closing R_8) leads to a dominant /b/ percept (represented by a square); increasing F_2 and F_3 (corresponding to closing R_6) leads to a dominant /d/ percept (diamond); increasing F_2 and decreasing F_3 (corresponding to closing R_5) leads to a dominant /g/ percept (triangle).

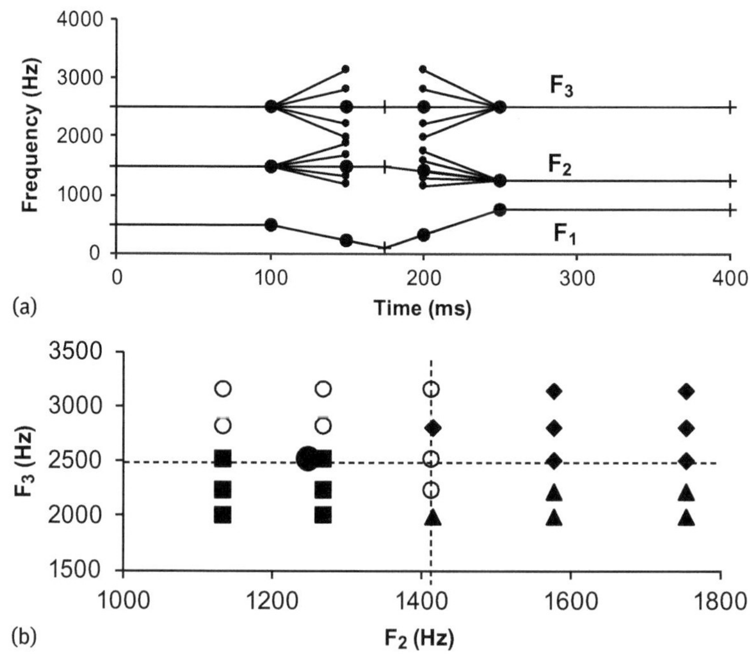

Fig. 7.18: Panel (a): Time-domain diagram of formant transitions of the stimulus for the perception of the consonant C in /əCa/ tokens (after Carré et al., 2002, Figures 2 and 3, p. 1682, with permission by the International Phonetic Association). Panel (b): /əCa/ results in the F_2-F_3 plane. For an explanation see Figure 7.17. The large circle represents the second vowel.

$F_3 = 2{,}500$ Hz to the boundary between R_5 and R_6 or between R_7 and R_8. A decrease of F_2 combined with an increase of F_3 (corresponding to R_7), led to a percept that could not be classified by the French listeners. The large circle represents F_2 and F_3 for the second vowel (corresponding in this case to the uniform tube). The orderly results of these perception tests suggest that as long as the direction of the transition fits one of the quadrants (ascending or descending vs. F_2 or F_3), the consonants perceived are exactly as those in the French language, that is, they correspond to the three plosive places of articulation in French. Since the responses show no crossover between quadrants, the results suggest that as long as the transition *directions* fit into the appropriate quadrant, they will generate that percept, regardless of the *slope* of the transition.

Other experiments (Carré et al., 2002) with vowels [əCa] and [aCa] also show the three main quadrants centered around the baseline [əəa] and [aəa] (Figure 7.18 and 7.19). The results suggest that perceptual identification of plosives can be carried out with reference to the neutral "schwa" in a very easy way – by simply look-

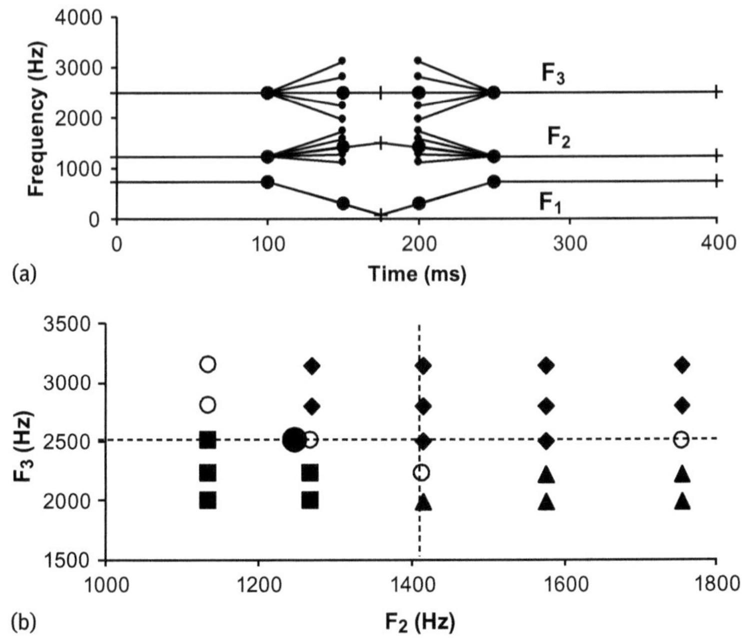

Fig. 7.19: Panel (a): Time-domain diagram of formant transitions of the stimulus for the perception of the consonant C in /aCa/ tokens (Carré et al., 2002, Figure 6, p. 1683, with permission by the International Phonetic Association). Panel (b): /aCa/ results in the F_2-F_3 plane. For an explanation see Figure 7.17.

ing at the rising or falling variation of the formant transitions with respect to the baseline. The baseline is the neutral or close to the neutral, or it could have one neutral transition extreme, depending on the degree of co-production. Although the actual values of the transition extremes are speaker-dependent, a listener can nevertheless identify the correct consonant from the rising and falling formant transitions alone (Ralston and Sawusch, 1984).

The use of the neutral position as a reference for coding or decoding transitional cues simplifies phonemic representation and identification. This position is completely specific in speech: it corresponds to the uniform tube at the production level. Vowel production is often characterized in terms of distortion starting from the neutral shape (see for example Harshman et al., 1977). CV_1 transitions can be obtained by superposition of C-neutral and neutral-V_1 trajectories (Mrayati et al., 1990; Carré and Mrayati, 1990).

This specific situation can also facilitate the process of identification of reduction situations. The special statute of the "schwa" in speech production allows us to understand why it can be used as a reference in speech perception, as evidenced in the present results: the simple detection of the falling or rising of the formant transitions is sufficient for categorization.

Finally, this implies that plosive consonants may be correctly identified irrespective of the phonetic context, as if every vowel were neutral.

The results of these last experiments on V_1CV_2 (with reference to the neutral) could seem contradictory to those on V_1V_2 (with transition rate measurements). But a deconvolution process could be hypothesized dividing V_1CV_2 into a V_1V_2 vowel transition and a C consonant around the neutral. More experiments must be undertaken to understand these phenomena better.

7.3 Discussion and conclusions

Findings presented in this chapter are consistent with the notion, as proposed, for example by Fowler (1980), that time, in the most general sense, is an important dimension of speech. Keeping this notion in mind, a few points need to be discussed.

The first of these compares two fundamentally different approaches: the once exclusive but still prevailing static approach with the dynamic one that the present book advocates. Although information carried by static data, such as the spectral representation of speech sounds, does not lack validity, it is both insufficient and, when used for the analysis of natural speech, also potentially misleading. Formant frequencies of vowels are intrinsic characteristics of the static, while they are extrinsic in the dynamic approach. The contrast between the two approaches extends to analysis methods as well: in the static approach they are based on frequency-domain measurements in time frames, whereas in the dynamic approach they are focused on frequency-domain changes that occur over time. Furthermore, normalization in the static approach generally takes place in the frequency domain, whereas in the dynamic approach it has to be performed in the time domain. Time-based normalization naturally accommodates various individual differences, such as speaker- or dialect-dependent transition or syllable durations. In addition, coarticulation phenomena should also be reconsidered from the viewpoint of temporal analysis of speech (Carré, 2011): it could be pictured as a type of deconvolution (Yu et al., 2007). Third, the preeminent role of transition rate in the perception of speech sound sequences should be emphasized: classic results that showed the manner of articulation in CV sequences as a function of transition rate (e.g., Miller, 1981) report only one of several phenomena dependent on the rate of formant frequency change, some of which also interact with speaking rate.

The above points lead us to proposing that the *invariance of transition rate* is stronger than the one of formant targets. Invariance of transition rate can explain, among others, results obtained by Strange et al. (1983) on the perception of C_1VC_2 tokens, in which a portion of the vowel is replaced by silence (silent center). Such operation does not affect the timing of syllabic production and the vowel can be identified based on the rates of transition. Strange showed that vowel identification of exemplars with such silent centers is better than the same portion of the vowel alone, that is in exemplars with the onset and coda transitions removed. These experiments also show the overriding importance of syllabic rhythm because, when the duration of the silent center is shortened while keeping the rates of formant transitions intact, the syllabic rhythm changes causing a decrease in the identifiability of the vowel (Verbrugge et al., 1976).

One shortcoming of the perceptual data presented in this chapter is that almost all of them were obtained with synthesized stimuli conforming to the French language and using French-speaking listeners, although some results did compare French and English listeners using French and English stimuli (Carré et al., 2001). Replicating the experiments with material proper to different languages and using native speakers of those languages as listeners would strengthen the validity of the data and of the conclusions based thereon. In addition, investigating the perception of transition rate in children would shed a light on the acquisition of the speech code over the developmental process.

From the results presented in this chapter and the discussion they inspired several conclusions may be drawn.

- Speech is a dynamic process characterized by kinetic movement of formant transitions: trajectory, direction, duration, speed, and acceleration in which time is a crucial dimension.
- Deformation gestures modeled by the DRM form a set of dynamic primitives of the speech production system, that is, they are identified as elements of the code for speech production.
- If gestures are indeed primitives of the production and perception of speech, then these gestures must be relatively independent from one another and could even be considered autonomous. The temporal unit of production is the syllable obtained by co-production of gestures.
- In the F_1-F_2 plane, vocalic transitions with duration controlled by the stable syllabic rhythm are distinguished by their different rates along distinctive trajectories of the DRM. Large changes in transition duration do not affect the perception of primitive gestures. The range of transition durations is speaker- and language-dependent. Duration is essential for identifying the transition as being consonantal, diphthong, or vowel-to-vowel. The degree of synchrony between two co-produced primitive gestures tolerates large phasing differ-

ences without changing the percept. Perception of a transition identifies the place of articulation and it is the result of a speaker-to-listener temporal adaptation process.

- Formant transitions on trajectories outside the vowel triangle perceived as vowel-to-vowel transitions indicate that, to specify vocalic transitions, information on their duration and direction is sufficient.
- Since the direction and the rate of transitions adequately explain vowel perception, this fact points to the necessity of reinterpreting normalization, coarticulation, vowel reduction, and especially invariance.
- Coarticulation can be regarded as the convolution of two gestures; by the same token perception of coarticulation suggests the presence of a deconvolution mechanism.
- In sum, dynamic properties of the DRM make it a tool exquisitely suited for the study of a speech communication system, a system that is fundamentally and intrinsically dynamic.

8 Speech perception viewed from the model

> *"... [We] believe that the only way to describe human speech perception is to describe not the perception itself but the artificial speech understanding system that is most compatible with the experimental data obtained in speech ... research."*
> *(Chistovich, 1980).*

In the previous chapters a model of speech perception was presented. The model was the outgrowth of a deductive approach starting from a simple acoustic tube having the length of the average male vocal tract and, when excited by a periodic source, the sounds generated were shown to have the same perceptual characteristics as those produced by natural speech. Thus, the above quote by Chistovich signifies that the auditory system is eminently capable of, if not outright specialized for, the perception of speech sounds, with high emphasis given to the most important property of the model: dynamics. In this chapter we will attempt to show that the processing of dynamically changing acoustic stimuli is, indeed, a primary – if not the primary – characteristic of auditory perception.

8.1 Properties of the auditory system, in a nutshell

It amounts to a commonplace statement to say that the auditory system is the sensory organ best suited for registering sound waves as well as changes in the acoustic environment with minimal delay. Indeed, the majority of vertebrates is equipped with ears that are capable of receiving and transmitting to their brain even very faint and brief clicks, noises, whistles, or other sound vibrations that have been generated by some abrupt or ongoing changes in the physical environment. Obviously, such capabilities must have developed for survival: to be alerted to noises made by prey, predators, or that of a likely mate. The environmental changes in question are events that generate or modify sound waves in the medium in which the vertebrate lives: air or water. The sound waves are thus produced by mechanical systems either directly, in response to application of some external force, or indirectly, by reinforcing certain resonant modes of sound waves produced by another source. Therefore, the vertebrate auditory system had to evolve to detect and recognize patterns of sounds produced by both types of mechanical systems. Also, the location of the event generating the sound had to be recognized: quite accurately in the horizontal plane, relying on binaural intensity level and time differences. The distance from a sound-generating event is estimated by the level and the presence of high-frequency components, both of which

DOI 10.1515/9781501502019-009

dominate in close sounds. Auditory location cues are very effective supplements to the more precise visual cues should they be unavailable due to low light or when objects are covered or ambiguous.

Most important of all is the fact that the human auditory system had to adapt itself to speech communication. As described in previous chapters, speech is a process that takes advantage of the presence of the vocal folds and the vocal tract – the first a direct and the second a resonating indirect sound-producing acoustic system. But is the auditory system of man optimally tuned to receiving and decoding speech sounds? Before addressing this question though, in the last subsection of this chapter, we would like to briefly show what the human auditory capabilities are: how do they respond to the diverse dimensions of simple and complex acoustic signals.

To begin with, the audible frequency range extends from about 50 Hz up to about 14 kHz for young individuals with normal hearing; although the high-frequency limit decreases with age, it still encloses the speech range of 50 to 5,000 Hz. The frequency scale that the almost eight-octave range of audible frequencies covers is remarkably fine. Indeed, the frequency of a tone can be discriminated from a higher or lower one when its difference amounts to no more than about 0.25% between 500 and 3,000 Hz, and about 1% at 6,000 Hz (Delhommeau et al., 2005). The intensity range between just audible and dangerously loud (not exceeding 120 dB re/ 2^5 N/m^2 with exposure shorter than 9 s) is 120 dB at 3 kHz but decreasing to about 55 dB at 50 Hz and 85 dB at 12 kHz. The sensitivity of the ear is also fine when it comes to sound level: the intensity of a tone can be discriminated from a louder or softer one at a difference of only about 1% (Hanna, 1986). However, the perceptual magnitude displays a compressive – almost logarithmic (dB-like) – non-linearity (Goldstein, 1967). The fine response characteristics of the ear reflect the precision observable at its most peripheral stage: on the basilar membrane (BM) and in the orderly rows of inner ear hair cells that the BM stimulates. Each point on the BM responds maximally to excitation by one frequency but, because of its physical characteristics, the excitation also spreads to neighboring regions. This spread generates a psychophysically observable excitation pattern (Scharf, 1964) – a measure of biophysical frequency selectivity that behaves like a bank of band-pass filters with asymmetrical side bands. The response profiles of these filters are propagated from the lowest (the basilar membrane in the inner ear) all the way up to the highest level of the auditory system (the auditory cortex). The frequency spreads can be observed in various situations using behavioral methods, whether as tone-by-noise masking patterns (Moore et al., 1990) or as psychoacoustic tuning curves (Johnson-Davies and Patterson, 1979). The spectral bandwidth of the excitation patterns, as well that of masking and tuning, displays a monotonic (although non-linear) increase on a Hertz scale. This frequency band gave rise early on to the notion of "critical bandwidth" (CB, Fletcher, 1940) as well as to the Bark scale (Zwicker, 1961). More recently, these bandwidth measures have been replaced by the physiologically more firmly based "equivalent rectangular bandwidth" (ERB) calculated by integrating the area between frequency limits corresponding to some fixed psychophysical or physiological performance level computed over the audible frequency range (Moore and Glasberg, 1983).

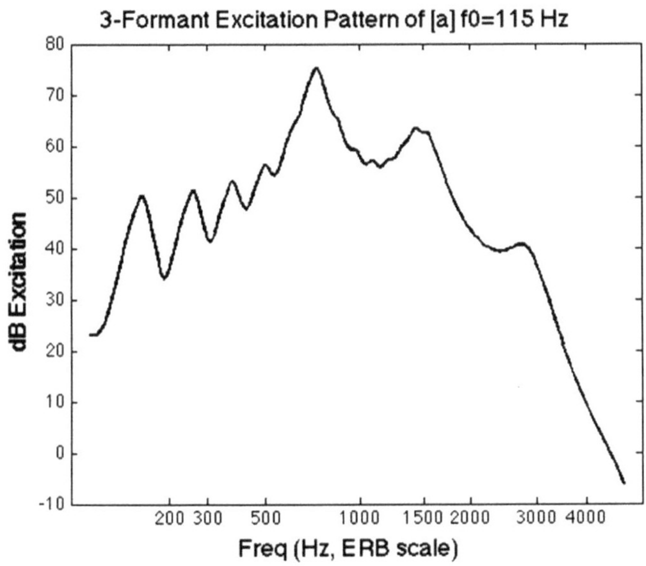

Fig. 8.1: Auditory excitation pattern of the first three formants of the vowel [a] spoken with a fundamental frequency of 115 Hz.

Sounds consisting of multiple simultaneous tones generate complex excitation patterns (Figure 8.1) that are often illustrated as a modified spectrogram reflecting the ear's non-linearity: compressed dB-magnitude on the ordinate and a linear-to-logarithmic low-to-high ERB frequency scale on the abscissa (Lyon's and Slaney's "cochleagram," Lyon, 1983, see Figure 8.2). Inputs under the same cochlear filter-ERB will influence each other: the more intense will mask the less intense and their loudness will sum. More generally, multiple inputs (tones and or band-limited noises) will generate auditory profiles (Green, 1988) – i.e., auditory patterns that, together with their individual temporal characteristics (duration, onset-offset, intensity increase and decrease) create different recognizable sound objects. When these complex objects contain only tones, two pattern classes are

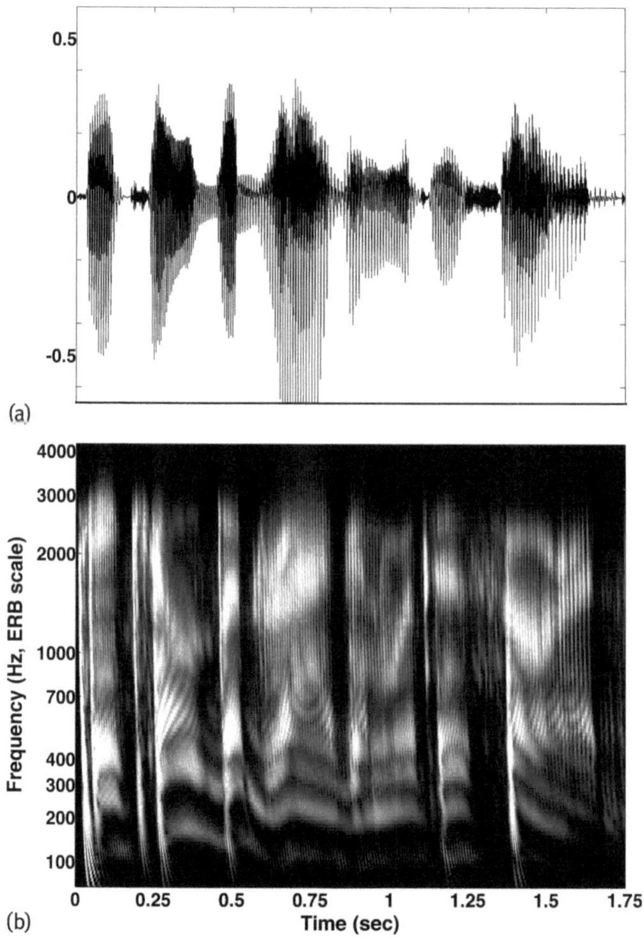

Fig. 8.2: Waveform [Panel (a)], and cochleagram [Panel (b)] of the sentence "He killed the dragon with his sword" spoken by a male talker.

distinguished: those with tones in inharmonic and those with tones in harmonic relationship.

While inharmonic patterns are less frequent in nature, harmonic patterns are abundant, whether in speech, music, or rhythmic environmental sounds. The latter are generated by periodic sounds, like pulse trains or non-sinusoidal waves, the spectrum of which contains components at frequencies that are multiples of the fundamental, that is of the frequency that is the inverse of the period's duration. The most salient perceptual feature of such sounds is *pitch* ("periodicity pitch," Schouten's, 1940, "residue pitch"), corresponding to the frequency of the fundamental (f_0). Interestingly, this pitch can be heard even when the component f_0 is removed and only its harmonics are present. The strength of the pitch percept is a function of the harmonic components present – the 3rd to the 6th creating the strongest salience – as well as their individual intensity; salience of the components themselves is determined by their relative intensity. Terhardt (1974) conducted extensive investigations on the combination of periodicity pitch and the pitch of the harmonic components. He called periodicity pitch "virtual" and that of the components "spectral." We hear pitch within the frequency range of ~40 to ~5,000 Hz. The spectrum of a periodic sound source is modulated by resonances of the body excited by the source: these resonances create the characteristic *timbre* of different musical instruments. The different *vowels* we produce when we continuously modify the shape of our vocal tract as we speak can also be looked upon as timbres. The spectrum of timbres shape the row of harmonics – lines at all multiples of f_0 – creating patterns of peaks and valleys that we have learned to attribute to a specific instrument, or speech sound (see Figure 8.2 b). Looking at the waveform of a sound that generates an audible pitch (Figure 8.2 a), we clearly see the periodic pattern.

Such sounds share properties with amplitude-modulated (AM) sounds generated by taking a steady sound (the carrier) and changing its amplitude by multiplying it with a periodic sound (the modulator); when the modulating frequency f_{mod} is higher than approximately 40 Hz, we also hear a pitch. AM sounds are everywhere in our acoustic environment. The auditory system is able to detect, recognize, and discriminate modulation frequencies; amplitude modulation changes can be well discriminated up to about 16 Hz (Demany and Semal, 1989), thereby allowing the ear to follow the flow of phonetic information even in consonant clusters like the one in the utterance "sixth street," while the best AM resolution, around 4 Hz (Dau et al., 1997; Fastl, 1982), coincides with the inverse of the average duration of a syllable (Greenberg, 1999), even across languages (Arai and Greenberg, 1997). The situation is somewhat similar when observing frequency modulation (FM): discrimination of FM rate (or, for rising or falling frequency sweeps, the slope of the sweep) is most accurate within the 2.5 to 30-Hz range, a range that includes the inverse of the syllable duration (making it possible to track the trajectory of formants between two successive vowels) and that of the consonant-vowel and vowel-consonant transitions (Kay and Matthews, 1947; Furukawa and Moore, 1997). While the up-or-down direction of FM sweeps is easily discriminated even at very short durations – 20 ms according to Schouten (1980), 50 ms according to Dooley and Moore (1988) – to accurately perceive its rate of change the auditory system seems to use the physical definition of velocity, that is the frequency sweep's extent divided by its duration, whenever both of these quantities are available to the listener (Pollack, 1968; Schouten, 1980). However, discrimination of the velocity of tone sweeps is less accurate but still possible even when their extent and duration are unknown, suggesting that the auditory system may possess a genuine velocity detector that focuses on the dynamics of frequency change (Pollack, 1968; Divenyi, 2005a; Crum and Hafter, 2008). The existence of such a detector would predict that the extent of the frequency change could be recovered from the perceived velocity whenever the duration and the onset frequency of the sweep are known or implied. Frequency sweeps are analogous to formant transitions in speech; they will be further discussed in the auditory dynamics subsection (8.4).

Although capabilities like the perception of pitch or of amplitude- and frequency-modulations do imply the existence of an internal timer that can measure the duration of pitch- or modulation-periods, auditory temporal detection and discrimination have been investigated also more directly. For example, the presence of a gap of only 2–3 ms can be detected in a burst of broad-band noise (Plomp, 1964), whereas in a band-limited noise the detection threshold is an inverse function of the bandwidth – about 35 ms at 50 Hz and about 5 ms at 1,500 Hz, showing a log-log progression (Eddins et al., 1992). Differences of 6 to 10% in the duration of static tones or noises are discriminable (Creelman, 1962) and so are empty intervals, that is when silent time intervals are marked by tone or noise pulses (Abel, 1972), but the just-noticeable differences increase as the frequency difference between the first and the second tone burst marker increases (Divenyi and Danner, 1977; Divenyi and Sachs, 1978), indicating that precise time perception is possible within, but not across, auditory frequency bands.

Auditory localization in the horizontal plane is performed by the auditory system taking advantage of the physical disparities of the sound waves as they reach the two ears, and locating the sound source based on interaural time and level differences (ITD and ILD). The location of a single source can be discriminated as finely as 1–3° at the center (= the "auditory fovea") and about 6–8° at the sides (Mills, 1958). However, localization of pairs of simultaneous sounds requires an angular distance of about 10° to more than 30°, depending on the relative similarity of the sounds (Divenyi and Oliver, 1989). Temporal synchrony of the right- and left-ear signals is one of the most important factors, especially for localization of AM signals. Room reverberation adversely affects localization, despite the presence of an echo-suppression mechanism in the human auditory system (Zurek, 1979). The most important center of the localization mechanism has been identified in the superior olivary complex of the brainstem that receives input from the right and the left sides; the medial nucleus is able to finely track interaural time delay, whereas the frequency analyzer in the lateral nucleus gauges interaural level differences across the spectrum. By the time the auditory information reaches the inferior colliculus, auditory localization has been established and is ready to be integrated with visual spatial information present in the superior colliculus. One can conclude that the localization system is efficiently organized across sensory domains.

We recognize that the above paragraphs give a woefully incomplete picture of the auditory system. To learn more about this topic, the reader is encouraged to get hold of a book on hearing, among those available the best researched and most popular being the one by Brian Moore (Moore, 2013).

8.2 Auditory pattern recognition

In our real acoustic world, simple sounds – sine waves, band-limited Gaussian (= random) noises – occur randomly; most auditory objects, "...distinct ...perceptual entities that depend on available brain mechanisms to represent and analyze [them]" (Griffiths and Warren, 2004), are complex spectro-temporal *patterns* – such as a middle-C note played on the piano, a cell-phone's ringing, the beginning four-note motive in orchestral unison of Beethoven's 5th Symphony, a dog's bark, a heavy truck passing by, a small child calling "mommy," the audience in a movie theater suddenly bursting into laughter, a large wave breaking on the seashore, etc. As we implied in the preceding paragraphs, the auditory system is capable of analyzing static as well as time-varying sounds, so mechanisms must exist that create perceptual representation of sounds no matter how complex – whether as auditory spectrograms, or pitch images, or spatial images – and form patterns that become objects to be perceived (or *aperceived*, as Wundt, 1874, said). But what kinds of patterns and how well do we recognize them?

If we think of musical patterns like notes played by different instruments, or chords, or tone sequences, ample data show the human ability to perceive differences within each class of patterns as well as to learn and identify them. For example, different instruments produce different timbres that can be discriminated based on their spectra as well as on the evolution of their spectra over time; listeners achieve differentiation of instrumental timbres by first learning to associate them with a given instrument and then classifying and scaling them along perceptual axes that form a multi-dimensional space remarkably similar across individuals (see Plomp, 1970; Grey, 1976).

In the acoustic domain, this classification process follows the density of harmonic components across the spectrum (Brown, 1999). Chords are classified principally along the consonance-dissonance continuum, which is, again, strongly tied to acoustic characteristics – this time to the degree of complexity of the ratio of the harmonic components: the simpler the ratio, the more consonant the chord (von Helmholtz, 1862, 1877; Roederer, 1975). Perception of tone sequences has been studied as the identification of the temporal order of two (Hirsh, 1959) or three tones (Divenyi and Hirsh, 1974, 1978), which showed increasing difficulty as the duration of the components decreased and as their frequency span increased. When a three-tone pattern is inserted in a longer sequence, identification requires longer durations in the middle, shorter at the beginning, and shortest at the end of the sequence. Asking the listener to notice a change in the frequency of a component of a pair of sequences consisting of tonal components randomly selected on each trial also produced best results at the beginning or at the end of the sequence, or when the component was an outlier in the spectrum, or (generally speaking) when the spectral or temporal uncertainty of the component was lowest (Watson et al., 1975, 1976). In summary, a tonal pattern is most accurately perceived when some of its attributes "stick out" – become more salient – compared to other attributes; salience of a component makes it easier to recognize the pattern. For static chords, salience is increased by dissonance, whereas in dynamic sequences salience of components is relatively high at the temporal and/or spectral edges of the sequence. Regarding the temporal processes behind the recognition of sequences, Ira Hirsh identified three, each tied to one particular time range: sequences of 20 ms and shorter are phase-related and are perceived as sound quality (plosive bursts and glottal stops in speech), those between 20 and about 150 ms become patterns and are perceived on the basis of dynamic frequency changes (like CV and VC diphones in speech), whereas those longer than 150 ms are perceived as consecutive separate perceptual entities (syllables in speech, Hirsh, 1974; Divenyi, 2004). Electro- and magneto-encephalographic (EEG and MEG) studies showed that these three behaviorally observed temporal ranges are also visible in brainwaves and are related to distinct electromagnetic frequency ranges in the brain – the syllabic range to theta, the shortest range to gamma, and the medium range to beta oscillations. Recognizable auditory patterns, including those of speech segments are consistent with the cognitive functions associated with beta waves (Ghitza and Greenberg, 2009; Giraud and Poeppel, 2012). Obviously, all three time ranges are important for the perception of the dynamically changing acoustic world around us; consequently, simultaneous reliance of the perceptual system on more than one brain oscillation frequency has been proposed by neurophysiologists (Poeppel, 2003).

Environmental sounds also represent a class of auditory patterns, the importance of which cannot be underestimated. Over the last 15 years considerable effort has been devoted to research on the classification, identification, and recognition of these sounds (Ballas, 1993; Gygi and Shapiro, 2010). Studies investigating the type and degree of expertise have pointed to the complexity of human recognition of environmental sounds (e.g., Lemaitre et al., 2010), while computational methods have been developed that would permit automatic recognition of a number of everyday sounds (Chu et al., 2009).

Turning to speech, prosody and individual voice characteristics do not carry phonological information but they do constitute patterns that need to be recognized and correctly interpreted by the listener. Parts of prosodic patterns are linguistic – such as the f_0 patterns in tone languages and differences between the same utterance being declarative or interrogative – but most of them are paralinguistic: they can convey an almost infinite number of emotional states (Frick, 1985). Pitch range and overall level, for example, could also convey friendly/unfriendly or happy/unhappy feelings (Ohala, 2010); pitch contours could indicate momentary or general emotional state, like anger, fright (Pollack et al., 1960), or even abnormal psychopathological status (Scherer, 1979). Introducing spectral changes, for example by shifting all formants upward or downward, or adding breathiness noise, also communicates emotion (Charfuelan and Schröder, 2011).

8.3 Multiple auditory objects: information and streams

In real life, an auditory object seldom appears alone: we are constantly surrounded by multiple, simultaneous sounds. Although, when we listen to a musical piece, the simultaneous presence of different instruments or voices are part of the experience and we enjoy following any or all of them at once, often we want to attend to only one of the sound sources and consider any others present as noise, interference, nuisance that we want to ignore or block out.

To better understand how the auditory system deals with multiple objects, let's consider the case of only two presented simultaneously. We have talked about masking: a sound will reduce or obliterate the audibility of a simultaneously present (or immediately following non-simultaneous) less intense sound in the frequency range where their responses overlap – this type of masking is called *energetic masking* (EM). However, one can be perfectly able to detect the presence of speech in the midst of helicopter noise but may not be able to understand what is being said, because it undergoes *informational masking* (IM). The latter has been defined as the difference between total masking and masking solely attributable to masking at the level of the peripheral auditory system (which is purely energetic, see Durlach et al., 2003). IM has also been quantitatively defined as the degree of *relative entropy* (a mathematical definition of dissimilarity) between two simultaneous sounds (Lutfi, 1993; Oh and Lutfi, 1999) as well as the degree of *uncertainty* of the information in the target probe, created by the presence of the masker (Watson, 2005). Since uncertainty in complex signals, such as sequences of random-frequency tones, can be reduced by both psychophysical and selective attentional training (Watson and Kidd, 2007), IM can be considered as the central component of total masking – i.e., the central *complement* of energetic masking. Thus, IM, just as EM, is likely to occur when two complex signals contain overlapping spectral regions in identical temporal regions. Masking of one AM signal – tone (Yost et al., 1989) or noise (Houtgast, 1989) – by another can only be informational when both signals are audible (Yost et al., 1989). The same is true for one FM signal masking another (Wilson et al., 1990). One could consider AM and FM masking as interference between patterns: amplitude modulation generates a pattern with regular envelope fluctuations and frequency modulation leads to a pattern of frequency fluctuations. When the difference of the same type of fluctuation between simultaneously present patterns is discriminable (see Section 7.1), one pattern can interfere with the perception of the other – in other words, simultaneous patterns can exert IM on one another.

One often-discussed example of IM is the so-called "cocktail party effect" (CPE) – a term proposed by Cherry (1953; 1966): listening to the speech of one particular talker over the background of other persons talking. Because of the nefarious effect of CPE on communication (especially for persons over about 60 years of age even without appreciable hearing loss (Divenyi and Haupt, 1997c, 1997b, 1997a), a considerable amount of research has been performed over the last 15–20 years, aimed at characterizing and quantitatively specifying the process. Thus, for example, we learned that increasing the spatial separation between the target speech and the interfering babble improves intelligibility by a degree corresponding to a signal-to-noise ratio (SNR) increase up to 10 dB (Bronkhorst, 2000). However, only a 7.5-dB increase was found when the SNR was measured separately in each critical band, arguing that IM is responsible for the largest portion of spatial advantage because band-by-band advantage (that is, the difference between SNRs measured at two locations within each band, akin to EM) predicts only a 1 to 3-dB effect (Shinn-Cunningham et al., 2005). Another important factor is the gender of the target talker and that of the interfering talkers. While at the largest spatial disparity, 0 vs. 90° separation, almost no performance difference was found between same-sex and different-sex target and interfering speakers (Festen and Plomp, 1990), in diotically presented conditions a 4 dB (four talkers) and 5 dB (three talkers) same/different-sex advantage was seen (Brungart et al., 2001), whereas in conditions with target and interferer presented directly to one ear at close distance (1 m) different-sex stimuli produced substantially better intelligibility (Brungart and Simpson, 2002). The number of interfering talkers of the same gender as the target talker produced only a moderate effect: a 3-dB SNR intelligibility decrement from four to ten non-target talkers, but knowing the target talker's voice characteristics (by way of priming) produced a 4-dB improvement (Freyman et al., 2004). In sum, in order to gain good understanding of what a target talker is saying in a CPE setting, it helps (1) to have a reasonably large spatial separation between target and interfering talkers, (2) to know the target talker, (3) to have all non-target talkers be of the gender different from that of the target, and (4) be younger than 60 years of age with at most a mild hearing loss. The considerable engineering and computational research effort in recent years aimed at developing a device that would help the user hear well in a CPE situation (e.g., Divenyi, 2005c; Wang and Brown, 2006) has been motivated by the recognition that the three above listed requisites for speech understanding in a speech interference setting are difficult to guarantee. But does a complex sound or sound sequence constitute a single object? This question was the driving force behind a whole line of investigations and theoretical studies concerning *auditory scene analysis* (ASA), best summarized in the classic book by Albert Bregman (Bregman, 1990). The brief answer is that it depends on the temporal and spectral structure of the sound, which can lead to it being perceived as a single *stream* or as an ensemble of two or several simultaneous streams. What forms a stream is the process of *grouping* spectral and/or temporal components of the sound that have common features. For example, harmonics of a common fundamental frequency will group together to form a stream – a given person's voice or a given instrument in a band – or common temporal envelope contours across frequency bands can form a stream and indicate a single source with a single amplitude modulation pattern – as in the acoustic output of a pneumatic hammer. One paradigm of stream fusion and stream segregation, widely used since van Noorden's (1975, 1977) paradigmatic study, is the continuous ...ABA_ABA_ABA... sequence where A and B are pure tones of identical duration but of different frequency. When the frequency difference between A and B is relatively small and the rate of the sequence is relatively slow, the sequence will form a single stream, whereas for larger frequency differences and faster rates the listener will hear two segregated streams, one as ...A_A_A_A... and one as (half as rapid)...B___B___B... Composers took advantage of this effect to obtain two independent melodic lines even on instruments that can produce only one (like the flute), by regularly alternating between two frequency ranges. Bach was a master of this technique and used it in his solo violin, cello, and keyboard pieces, like the one illustrated in Figure 8.3 taken from his Toccata and Fugue in d-minor written for organ. Our experiments showed that perception of time intervals (Divenyi and Danner, 1977) or the temporal order of tones (Divenyi and Hirsh, 1978) is good within, but loses precision across streams. The same is true for the stream fission limit of ABA_ABA_ABA... sequences using for A and B either harmonic (Grimault et al., 2000) or inharmonic (Valentine and Lentz, 2008) tone complexes: wider separations of the fundamentals of A and B result in fission. Actually, discriminability of a time delay introduced before the B sounds decreases when the spectral difference between A and B is large enough to

Fig. 8.3: Fugue theme from J. S. Bach's Toccata and Fugue in d minor for organ (BWV 565). The series of "a" notes constitute one stream (arrow 1) and the rest the other stream (starting with a "g" and ending with an "f" note, arrow 2).

make the sequence split into two streams (Vliegen and Oxenham, 1999). In other words, at wider frequency separations, tones cannot be integrated into a single stream and the auditory system is capable of precise temporal processing only within, but not across, streams. Consciously or unconsciously, composers and improvisers use this phenomenon to create a single or several interleaved musical streams and people have learned to communicate by trying to achieve maximum segregation between the voices of talkers speaking at once. And they do that by focusing on segregating *vowels* (de Cheveigné et al., 1995).

8.4 The dynamics of the auditory system

Looking back at the preceding sections of this chapter, the message emerges that the auditory system is optimally suited for the perception of a special type of auditory pattern: speech. We have also talked about auditory dynamics – the processing of amplitude- and frequency modulations and the perception of frequency glides. In the present section we want to go a step further because, as the previous chapters argued, we consider speech as an essentially dynamic process in which the information is embodied in the changes. Thus, we regard perception of speech not as that of separate elements (i.e., as phoneme-by-phoneme perception) but in terms of transitions from one point in the speech stream to another, where the point of changing transition direction (or velocity) may have a phonetic label attached to it but, for the auditory system's purpose, it is just a discontinuity: a singularity in a continuous function. The importance of transitions in the perception of nonsense CVC syllables has been shown by studies, in which either the vowel or the consonants (Strange, Verbrugge, Shankweiler, et al., 1976; Kewley-Port et al., 2007; Fogerty and Kewley-Port, 2009; Fogerty and Humes, 2010) were replaced by silent periods or by speech-spectrum noise of identical duration. As long as the vowel nuclei are present – inevitably bracketed by vestiges of transitions preceding and following the vowel nucleus – even these severely truncated CV and VC transitions will make speech understanding possible. As Fogerty and colleagues conclude, "...vowel information is essential for [word and] sentence intelligibility" and "... vowels provide a two-to-one benefit over consonants for the intelligibility of sentences" (Fogerty and Kewley-Port, 2009; Fogerty and Humes, 2010).

Among the first to recognize the importance of the dynamics of speech production and perception were Kozhevnikov and Chistovich at the Pavlov Institute in Leningrad. The principal contribution of their pioneering studies, condensed in a currently hard-to-find monograph (Kozhevnikov and Chistovich, 1965), was to show the link between articulatory and acoustic dynamics by way of physiological and behavioral measurements, and by building a solid theoretical basis to account for their data. Chistovich was particularly interested in the acoustics and the perception of formant transitions. As a conclusion of her investigation on the influence of both transition extent and transition rate on the perception of CV and VC syllables, she proposed that the auditory system must have two low-level subsystems for the analysis of speech, a peripheral one for spectral analysis and a more centrally located one for the analysis of changes in the spectral patterns (Chistovich, Lublinskaya, et al., 1982). She postulated that the second subsystem performed an online process that classified the input into previously learned categories. In other words, her theory extended into perceptual dynamics the motor theory of Liberman and Cooper that was based on categorical perception results (Liberman et al., 1954, 1967); moreover, it also foreshadowed neural network classifiers used over the last 30 years in research on automatic speech recognition (e.g., Lippmann, 1987; Dahl et al., 2012).

The work presented in the previous chapters to a large extent benefited from, and confirmed, Chistovich's studies. As an example, Chistovich (1980) demonstrated that, although a simple shift in the frequency of F_2 led to changing an /i/ vowel into /u/, the same parametric modification was insufficient to perceive a shift from /i/ to /U/, where an F_1 change was also necessary. While, on the surface, this finding may look modest, it underscores the overwhelming importance of formant trajectories as the major perceptual cue carrying vowel identity information – a concept that also represents one of the foundations of our model and serves as the starting point in the experiments we undertook to substantiate it. In a similar vein, F_3 trajectories define the place of articulation cue in CV syllables, shown a long time ago at the Haskins Laboratory (Harris et al., 1958), and this result, too, has been replicated by our model: for a closure specified by the distinctive region corresponding to a given place of articulation, the F_2 and F_3 trajectories of the DRM are in good agreement with those found for the perception of CV syllables (see Chapter 7). In addition, our model also naturally incorporates two variables of CV and VC dynamics notoriously difficult to measure in natural speech: the onset and the offset delays of the transitions – points in time that represent measures of CV and VC coarticulation and indispensable for the specification of locus equations (Krull, 1989, see Section 5.5).

Thus, a wide range of observations points to the primary importance of formant transitions in the production of speech. However, as seen in the previous chapter, our model goes much further: it is built on the principle that information in speech resides in the dynamic changes themselves. So, the question to ask is whether there exists an auditory mechanism the sole role of which is to follow frequency changes, in general, and formant variations, in particular. In Section 8.1, we pointed out that the monotonic change in formant frequencies (a change that is linear on an auditory frequency scale, like the ERB) reflects the shift (i.e., a glide) that occurs in the resonant frequencies of the cavities that produce them when the shape of the tube is changed. Although this shift represents a change in the phase spectrum of the filter (the resonator), the mechanism responding to it must perform a task different from the mechanism that decodes periodic frequency modulation (FM). Consequently, monotonic unidirectional and periodic up-down frequency changes may involve different mechanisms that account for their perception. Correct identification of intended sweep targets, as in hypo-speech understanding (Lindblom, 1990b), vowel reduction (Divenyi, 2009, see Figure 8.4), masking of abruptly terminated tone glides at a frequency beyond those at which the glide stops (Kluender and Jenison, 1992; Crum and Hafter, 2008, see Figure 8.5), or listeners' perception of the continuity of tone glides interrupted either by a gap and a short opposite-direction glide in the gap (Kuroda et al., 2012, see Figure 8.6) or by a broadband noise burst (Ciocca and Bregman, 1987) are phenomena difficult to explain without postulating a mechanism that gauges the velocity of frequency change. Naturally, FM sweeps involving a single sine wave are not identical to formant glides encountered in speech. Most studies that have investigated discrimination of formant frequency sweeps have looked at their phonological effect, that is, their effect on the speech sound perceived. However, there have been a few investigations that measured the discriminability of formant sweeps by asking listeners to compare sweeps of different velocities and durations (Elliott et al., 1989; Ainsworth and Carré, 1997; Ainsworth, 1999). The ensemble of findings on the perception of frequency transitions can be explained by postulating the existence of a sweep-velocity evaluator mechansim[1].

Studies focused on the perception of glides – monotonic frequency changes – have revealed an interesting perceptual property, one that is part of the perceptual principles listed by early Gestalt psychologists (e.g., Koffka, 1935): the principle of

1 We disagree with Schouten's and Pols' (Schouten and Pols, 1989) conclusion that formant sweeps are essentially perceived from their starting and ending frequency, to which the transition rate and duration add no information. To make that conclusion, they should have first assessed the comparison of sweeps from different starting and final frequencies within each single trial, as Pollack (1968) and Divenyi (2005a) did in their sinewave glide experiments.

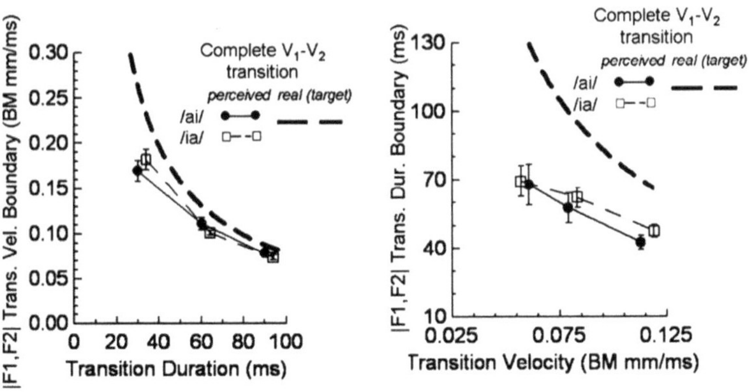

Fig. 8.4: Perception of the completeness of an incomplete V_1V_2 transition (from Divenyi, 2009, Figures 3a and 4a, pp. 1432–1433, with permission by the Acoustical Society of America). See Figure 7.13 for schematic time-frequency diagrams of the stimuli illustrating the /ai/ transition trajectories of the first two formants in four tokens in each experimental condition. Left panel: results from the constant transition duration/different transition velocities experiment; on the abscissa: transition duration; on the ordinate: transition velocity boundary, that is the lowest velocity at which the listener perceived a complete V_1-V_2 trajectory. Right panel: results from the constant transition velocity/different transition durations experiment; on the abscissa: transition velocity; on the ordinate: transition duration corresponding to the VR boundary, that is the shortest duration at which the listener perceived a complete V_1-V_2 trajectory. The dashed line in each panel shows transition velocities across all durations at which the V_1-V_2 trajectory is complete.

continuity. Indeed, it has been shown that when one inserts a short gap in a long tone glide and a short tone glide of opposite direction is added during the gap, listeners report hearing that the long glide is continuous and the short glide has a gap (Kuroda et al., 2012, see Figure 8.6). Another phenomenon implying a continuity-seeking mechanism was reported in the backward masking of a long tone glide by a short narrow-band noise burst, observing a masking maximum when the center frequency of the noise was at the frequency the glide would have arrived at, had it continued into the noise burst, rather than at its actual final frequency where auditory frequency analysis would have predicted it to happen (see Crum and Hafter, 2008, see Figure 8.5). Both phenomena predict the existence of a frequency-change velocity detector that provides an ongoing estimate of the path of an input dynamically changing in the frequency domain.

The same Gestalt principles seem to play a role in the perception of vowels in a dynamic context. First, there is ample evidence that vowels in fluent speech are not static but that their formant frequencies display a mostly shallow U- (or upside-down U-) shaped contour. This general finding has been labeled "vowel-

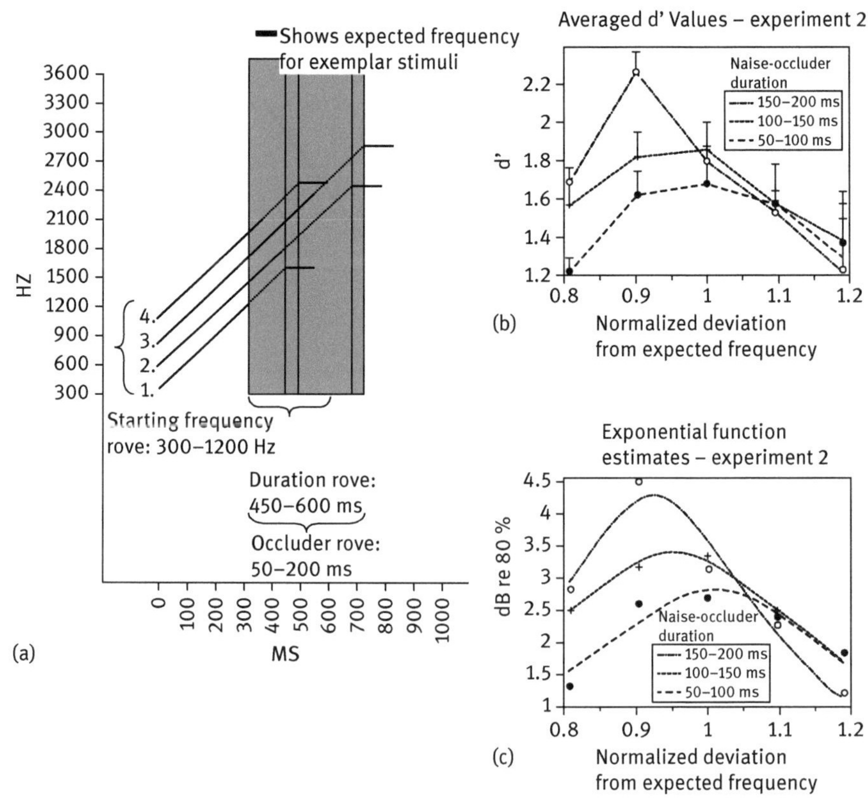

Fig. 8.5: Detection of a probe tone having a frequency following that of a tone glide followed by a broadband noise burst, with time-frequency structure shown in Panel (a). Both simultaneous and forward masking threshold of a tone of known or expected frequency is lower than that of a tone of random frequency. Panels (b) (detectability index d') and (c) (estimated probe threshold level), both as a function of tone frequency, show that the expected tone frequency was at (because of the after-offset continuation of the glide in the auditory nervous system) or slightly below (because of the time course of the growth of masking from low to high frequencies) the actual frequency. After Crum and Hafter (2008, Figures 3 and 4, pp. 1129 and 1132, with permission by the Acoustical Society of America).

inherent spectral change" (Morrison and Assmann, 2013). In particular, Lindblom observed that formants of vowels during carefully articulated "hyper-speech" reach values more peripheral (i.e., extreme) inside the vowel triangle than those during more casual "hypo-speech" (Lindblom et al., 1996). Nevertheless, while listeners do take note of the difference between the two types, the intelligibility of the two are identical, except under low SNR conditions. This realization prompted Lindblom to investigate the perception of incomplete vowel trajectories, which

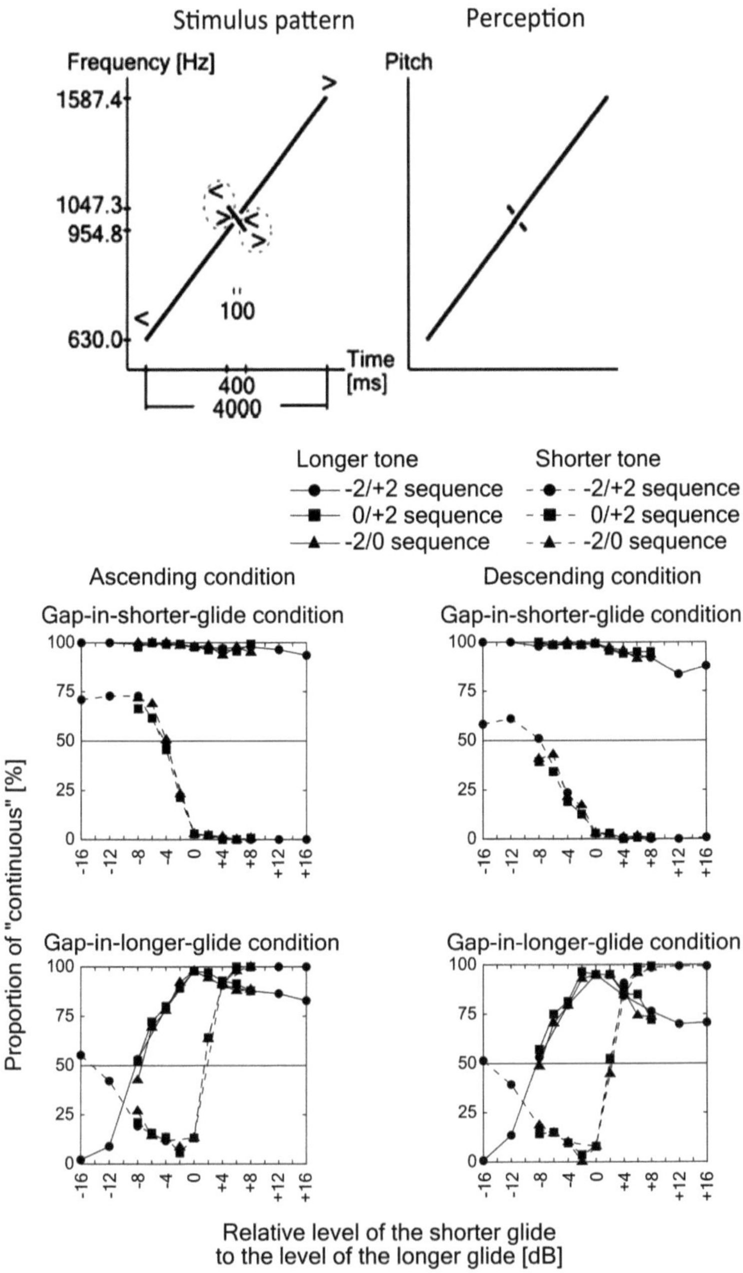

Fig. 8.6: Perception of continuity of a longer ascending tone glide interrupted by a shorter descending glide. The top panel is a time-frequency diagram of the stimulus for the ascending long tone glide. The bottom panels illustrate the proportion of "continuous" responses of both the longer (dark lines) and the shorter (light lines) glide as a function of the two glides' relative level, for both the ascending and the descending long glides. (From Kuroda et al., 2012, Figures 1 & 7, pp. 1255 & 1260, with permission by the American Psychological Association).

he labeled "vowel reduction" (VR) (VR, Lindblom, 1963). Although not without controversy, VR has been accepted as a characteristic of speech production. In fact, several experiments have shown that the almost-generally observable "undershoot" of the gesture into the vowel is not only accepted by the listeners but it is often not even perceived – it is compensated by a perceptual "overshoot" much like Crum's tone glide masking experiments (Crum and Hafter, 2008) would predict. Our studies (Divenyi, 2009, see Figure 8.4) have demonstrated a trade-off between transition duration and transition rate for V_1V_2 pairs resulting in VR (i.e., the transitions perceived as if they had reached the target the listener was primed with), constant within a given extent of V_1-V_2 difference and, as shown in Chapter 7, even outside the vowel triangle. Moreover, our experiments showed that a reduction also occurs in the perception of tone glides (Divenyi, 2009) pointing to the likelihood that the existence of a velocity tracking mechanism can provide a physiologically anchored explanation for not only VR but also for vowel perception outside the vowel triangle, as demonstrated in Subsection 7.2.1.3 of the preceding chapter. Incidentally, such a mechanism is necessary to explain results on the discrimination of the velocity of glides when information on both frequency extent and duration of the glides is severely curtailed by randomization (Pollack, 1968; Divenyi, 2005a).

As to the mechanisms involved, the spikes observed in auditory nerve fibers are statistically synchronized by the time domain shape of the basilar membrane excitation around the characteristic frequencies (Sachs et al., 1982) and may represent peripheral mechanisms underlying the perception of formant transitions. The spike trains can give information not only on the amplitudes of spectral components but also on the shape in the time domain of the components and thus on their phases. The phase variation (−180° around formant frequencies for second-order filters describing the transfer function of the vocal tract, see Fant, 1960), in fact, provides the mathematical basis underlying the transitions and their rate and could be used to measure the rate of transitions. Experiments with synthetic signals characterized by a flat amplitude spectrum and specific monotonic phase changes show that the changes are perceived as glides (Schroeder and Strube, 1986).

In non-speech perception experiments with sequences of tones having a glide as a first portion followed by a steady state portion, Ralston's (1984) listeners reliably sorted the sequences into three categories, corresponding roughly to rising, flat, and falling transitions. Although it is not to say that speech perception is entirely reducible to psychoacoustics, it is possible that such tone-and-glide categories are similar to the primitives needed for place of articulation distinction in speech and may have been used as a cue for perceiving place of articulation changes around the schwa position in the experiments described in Subsection 7.2.2.

However, most importantly for communication, speech dynamics involve changes in the temporal envelope – syllabic and sub-syllabic AM fluctuations. Syllables represent the shortest segments of speech that are both linguistic and paralinguistic – prosodic – and that can be defined in a simple and (relatively) uncontroversial manner (Arai and Greenberg, 1997). Experienced and unsophisticated listeners are equally able to parse (i.e., segment) into syllables both natural and even compressed speech (Wingfield, 1975). Conversely, reducing or smearing syllabic envelope fluctuations interferes with intelligibility (Drullman et al., 1994a, 1994b). Perceptual segmentation of speech into syllables, however, is not automatic. As to where the boundary between syllables is located, the answer is language-dependent and, even within a given language, precise specification of the boundary is difficult. The principal cue of syllabic segmentation appears to be the amplitude envelope either across the entire spectrum or within selected frequency bands (Ghitza, 1993). Syllable boundaries are conveyed almost all the time by minima in the envelope of the speech signal but they are also much influenced by prosody: for instance, English listeners are able to segment speech at the onset of strong, but not at the onset of weak syllables (Cutler and Norris, 1988).

In the perceptual realm, investigations have mainly been focused on syllabic and word-level segmentation (Turk and Shattuck-Hufnagel, 2000). As to sub-syllabic segmentation, whether by the listener or by the machine, the process is far more complicated: such a task involves inserting temporal segment marks inside a syllable between vowel and consonant or between consonants inside a cluster. According to Fant (1976), phonemic segmentation is not only difficult but also pointless, since it entails separating co-produced movements – an attempt contrary to foundations of articulatory phonology. We have mentioned locus as the formal representation of consonant onset and locus equation as that of the temporal flow of CV and VC segments (see Subsections 1.2.3.3 and 5.5). While these concepts succeed in defining analogs of segment marks, they are measured from recorded spoken tokens and they differ from one speaker to the next. Also, such time marks are post-hoc and do not pretend to assume that the perceptual system measures and uses the same temporal marks. To our knowledge no experimental measurements of the perception of phonemic segmentation marks have been attempted, with one exception: another experiment by Chistovich and her colleagues (Chistovich et al., 1975). In this experiment, the subject had to verbally produce [aC_2a] responses in synchrony with a repeatedly, but not regularly, presented [aC_1a] stimulus. The recorded stimulus and response traces indicated that the subject was able to obtain a perceptual time estimate for both the onset and the offset of the consonant in the stimulus. From these time values a perceptual duration of the consonant segment of the stimulus could be estimated and found to be around 60 to 70 ms on the average. After performing a follow-up experiment that used stimuli of pulse trains composed of different number and duration of identical frequency tones, Chistovich concluded that sequences of simple pulses were also segmented by the subject and that the minimum duration of the segments perceived was also around 60 to 70 ms[2]. This time range is inside the bracket of the one that Hirsh (1974) associates with the duration of auditory patterns – right between the shorter 2 to 30-ms span of phase-guided temporal analysis and the longer 100 to 200-ms span that signals separate auditory events. As we have shown in previous chapters, successions of any number of consonants and vowels, such as CV, VC, V_1V_2, $V_1C_1C_2(C_3\ldots)V_2$ sequences, involve rapid spectral changes – transitions that shift energy from one spectral band to another – and each of these changes contain a potential boundary on the time line that signals a particular property of the change. For example, as the above example by Chistovich shows, a boundary can appear at the onset and/or the offset of a CV or VC transition, just as the locus and the locus equation also specify. Or, as an alternative, one could define a boundary as a singularity of the formant frequency/time function (i.e., a point in time at which the formant change is either maximum or zero). While there are a number of algorithms and models that will automatically define sub-syllabic segments and determine segment marks (e.g., van Son and van Santen, 2005), boundaries between consonants and vowels, as well as the durations of sub-syllabic segments, remain difficult to define, both perceptually and computationally, because of coarticulation and co-production of the component gestures each of which follows its own dynamics.

Sub-syllabic and syllabic segmentation may be also viewed as representing different levels of prosodic dynamics (Byrd and Saltzman, 1998). One effect of prosody is that segments can be defined with higher accuracy for stressed than for unstressed syllables (Lehiste and Peterson, 1959; Sluijter and van Heuven, 1996).

8.5 Is auditory perception optimal for speech?

In the previous chapters we have characterized speech as a process producing gestures, that is, dynamic changes in the shape of a tube – the vocal tract. These

2 In a paper contemporary to Chistovich's study, the minimum duration for a chain of four identical-duration vowels was found to be 130 ms for diphthongized and 180 ms for isolated vowels (Dorman et al., 1975).

changes were primitives of a communication system: they were optimal for generating messages that the talker wanted to pass on to a listener. We have also showed that the dynamic changes of the tube's shape translated into dynamic changes of the natural resonant frequencies of the tube, thereby generating an acoustic signal continuously changing in its amplitude and its spectrum – in precise accordance with Rainer Plomp's much-quoted definition of speech (Plomp, 1983). While the previous subsections argued that the auditory system performs all kinds of tasks other than responding to and decoding speech – just as the mouth and the larynx, besides producing speech, are also engaged in many other activities – the question arises whether the perception of speech represents that one to which the auditory system is optimally tuned. One generally endorsed observation may give a positive answer to this question: speech perception is *robust*: we can understand it in noise, in badly reverberant spaces, sped up or slowed down to enormous degrees, with chunks of it obliterated, with its spectrum transposed up or down. No other type of acoustic signal is as resistant to abuse as speech is, when its perception is at stake. Therefore, although arriving at the conclusion from inductive observations rather than from deductive reasoning as we did when we concluded that the human speech production system was optimal, we are inclined to state that the auditory system is optimal for speech perception. But what does the auditory system need in order for the listener to optimally perceive the signal, even under less-than-optimal acoustic conditions? Obviously, it needs to be able to follow the spectral and temporal fluctuations of the signal, sample it, compare the sample pattern to previously encountered (and learned) patterns, and identify the one just heard as one in his/her learned repertoire in the particular language she/he is listening to.

Let's say that the particular sampled pattern is a vowel and that the language of communication has the [aiuɛɔeo] vowel system with seven vowels. Thus, for this language vowel perception becomes a 1-of-n ($n = 7$) identification task that can be reduced to a signal detection problem – extremely well treated by Macmillan and Creelman in their user's book on signal detection theory (Macmillan and Creelman, 1991). However, since the listener must understand the message spoken by the talker, he wants to be practically 100% certain to correctly identify the vowel sample in question. For this to happen, the distance between any pair of the seven-vowel set should be large enough to guarantee identification performance at a level of $d' = 3.0$ or higher. Specifically, we want to know precisely how large of a formant frequency difference can be discriminated at such a high performance level. Earlier in this chapter we mentioned studies on spectral profile analysis (Green, 1988) that consists of a task of comparing a multi-component signal to a similar one that differed only in the loudness of a given component. Studies on the discrimination of formant changes in vowels did much the same, by first evaluating the specific loudness (i.e., the bandwidth-weighted loudness Sone/Bark[3] as a function of frequency, see Moore and Glasberg, 1987) of each harmonic component of the vowel over the spectral region of a given formant, and then determining the amount of change in the level of that component discriminable at threshold (Kewley-Port and Zheng, 1998a). Since threshold in psychophysics almost always refers to a performance level of $d' = 1.0$, these threshold values had to be extrapolated to estimate formant differences yielding a $d' = 3.0$ performance. After such extrapolation we found that distances even between the first or the second formant of [e] and [ɛ] exceed the specific loudness differences estimated to yield a $d' = 3.0$ discriminability. These thresholds, therefore, provide psychophysical realism to the requirement that any two vowels in a given vowel system must be "sufficiently different" from one another, as proposed by Lindblom in his adaptive dispersion theory (Lindblom, 1986a), a theory also drawn on estimates of the specific loudness of vowels.

Although the data on the discrimination of formant changes as well as the adaptive dispersion theory are based on the comparison of two static vowel samples, they purport important implications for our dynamic tube model. As shown in Chapter 3, applying the criterion of minimum effort for maximum acoustic change will identify natural formant trajectories up and down to the limits of the vowel triangle when its output is controlled only by the tube's inherent limitations (see Figures 3.5 and 3.6). However, the presence of intermediate vowels on the trajectory primitives, like the vowel [e] on the [a-i] trajectory, indicates that the evolution of these trajectories had to be controlled by a mechanism capable of precisely gauging the dynamics that led from a given starting vowel to a "sufficiently different" other vowel. We said *dynamics*, first of all because the temporal succession of two vowels suggests a rhythm of consecutive syllables, a rhythm of 4 to 5 Hz on the average and which, as explained in detail in Chapter 7, is the basic time scale of speech corresponding to the frequency at which jaw movements require the least effort (see introduction to Chapter 7 and Figure 7.3). The control mechanism, first, must have precise knowledge of the formant trajectory on which the transition from the first to the second vowel lies, as well as of at least two of the following three parameters: the "sufficiently different" second vowel's formant acoustics (or at least the value of one of its formants, since the formant trajectory is known), the velocity of the transition, and the duration of the transition. Since the duration is most likely to be around 250 ms, i.e., that of a syllable, knowing

---

3 Sone is the unit of subjective loudness, as defined by S. S. Stevens (1936), the Bark scale is a frequency scale proportional to the width of the Critical Band at each frequency, as proposed by E. Zwicker (1961).

the transition velocity will lead to the correct formant values of the second vowel, while knowing these formant values will specify the transition speed needed to get there in about 250 ms. We propose here that this control mechanism is the auditory system, for the simple reason that, as explained in the previous sections, it is more than appropriately equipped to assess (and remember) formant frequencies and formant frequency differences, durations, and frequency- and formant-shifts. Moreover, these capabilities have been shown to be optimal in the parametric ranges corresponding to these variables: 4-Hz fluctuations at which AM resolution is best (see Section 8.1 and Fastl, 1982), formant changes in the F_1 but especially in the F_2 regions (Kewley-Port and Zheng, 1998b), formant transition velocity corresponding to durations of 60–70 ms (Chistovich et al., 1975). It is to be understood that these capabilities belong to a dynamic system and, as such, it forms the second, the reception/perception half of the yin-yang of speech communication, the first half of which being the dynamic production system explained and modeled in Chapters 2 through 7. Its frequency-resolving capability is able to resolve formant peaks even when they are relatively close (such as F_1 and F_2 of /u/ or F_2 and F_3 of /i/); this and its exquisite sensitivity to respond essentially immediately to spectral changes makes it well suited for identifying places of articulation, in addition to all vowel features. In stop consonant-vowel CV and VC sequences for example, the up-or-down direction of the formant transition serves as the main cue for defining the consonant's place of articulation. This cue is very effectively perceived even at durations as short as 10 ms (Schouten and Pols, 1989). Also, the fine temporal resolution (less than 10%, Divenyi and Danner, 1977) of the ear makes it extremely well suited for identifying voicing and manner of articulation; its sensitivity to level changes across the frequency range makes it ideal to detect features like nasal, lateral, or affricate. As said in Section 8.1, the sensitivity of auditory time perception is especially sharp in three temporal ranges, the "Hirsh zones" (Hirsh, 1974; Divenyi, 2004): 2–30 ms (best to signal minute acoustic changes), 60–100 ms (ideal for perceiving auditory patterns), and 150–250 ms (for syllabic rhythm). As we stated in Section 8.2, these three ranges also correspond to specific brain wave frequencies: to the gamma, the beta, and the theta oscillations, respectively. In other words, the auditory system can act as a speech-to-text transcriber that, under relatively high signal-to-noise ratio (6 dB or better) acoustic conditions, could operate essentially error-free – a few percentage points over what the most successful automatic speech recognizer did in 2015.

In sum, characteristics and capabilities of the auditory system appear to be ideal for the perception of speech. Obviously, there must be a more central system that processes, learns, and predicts speech input to the ears. The learning process itself is a fascinating journey, starting with infants in the third or fourth month after birth being tested and discovered that they learn, remember, and retrieve phonetic information including dynamic changes, like formant glides necessary to perceive the manner of articulation feature (Hillenbrand et al., 1979). Thus, the ease to respond to dynamic patterns of speech may indicate genetic predisposition but learning these patterns early on biases the auditory system toward listening to speech and creating a preferential place for the parametric ranges or values important for speech processing, as listed above.

But speech consists of much more than the "what" of the message of the talker. For one, auditory localization is able to give the listener fairly accurate information on the talker's location in the horizontal plane and, under most normal circumstances, also the talker's distance from the listener. But the auditory system's exquisite sensitivity makes it also eminently suited for decoding most non-linguistic features lumped under the wings of prosody: identification and learning a speaker's individual and often idiosyncratic voice characteristics, recognition of his or her emotional state, recognition of the social setting in which the talker is speaking, recognition of a dialect or a specific second-language accent, recognition of acute or chronic voice or other speaking disorders, etc. All these abilities derive from the sensitivity of the auditory system that goes way beyond what is necessary for reception of the message. To be more precise, detection and discrimination abilities of the auditory system, at a $d' = 1.0$ threshold level, for frequency, spectral changes, pitch, amplitude and phase fluctuations, amplitude and frequency modulations, and changes in the (horizontal) spatial are fine enough to analyze the incoming speech in short-duration time frames and notice practically immediately any change along any of these acoustic dimensions.

One area in which such high-precision non-linguistic analysis of speech received is particularly helpful happens when the speech the listener wants to attend to is embedded in an interfering mass of extraneous sounds. If the interference is random noise – random-amplitude components all along the frequency scale – the amplitude fluctuations of the noise will necessarily include time periods during which its energy will be lower than that of the speech signal, making it possible to "listen in the troughs" (Buus, 1985). Of course, in random noise these fluctuations do not need to cover the whole spectrum; they can be confined to spectral bands where the formants of the speech signal are located. Thus, there will be periods during which one or several formants and formant transitions will be audible to the listener, creating time-frequency tiles over the speech-plus-noise spectrogram that will be dominated by the speech. A listener familiar with the talker's language is able to guess all or part of the message, depending on the signal-to-noise ratio (SNR) and on the degree of the listener's familiarity with the voice of the talker (Johnsrude et al., 2013) or with the topic of the message (Dekerle et al., 2014). Explaining these effects would make us enter the domain of selective attention – a process much involved in decoding the complex issue of the cocktail party effect (CPE, Cherry and Taylor, 1954) briefly described in Section 8.3. However, descending into the more down-to-earth zone of phonetics, the noise in one formant region may affect only a portion of the transition and, if the information let through by the noise is sufficient to identify the trajectory, chances are that the time period containing the particular CV or VC segment will allow that particular syllable (or at least half-syllable) to be correctly perceived. In other words, the listener's innate or acquired knowledge of trajectory primitives, paired with the unbroken presence of the syllabic metronome rhythm, is likely to provide a sufficiently solid frame on top of which the dynamic flow of speech can be reconstituted even from fractional information about the actual transitions. If using such a perceptual counterpart of our dynamic model makes it possible to actually extend (along the trajectories) the speech spectrogram's time-frequency tiles that troughs in the noise have allowed to be identified, one should ask the question what would happen with the tiles dominated by the noise, i.e., those in which the speech signal is masked. Extensive work on speech-noise separation by machines (e.g., Roweis, 2000; Wang, 2004) has shown that intelligibility of speech in noise can be vastly improved when the noise-dominated time-frequency tiles are simply deleted, by a process called "binary masking." Since performance by human listeners has also been shown (Kjems et al., 2009) to improve when binary masking is applied to noisy speech, we can hypothesize that the auditory system is capable of ignoring those time-frequency regions in which there is no speech information.

More difficult, both for the theories and the listener, is the CPE situation in which the interference is also speech: the "true" cocktail party setting where the poor listener is trying to understand what one particular talker is saying when dozens of others, left and right, are also talking. Here, the listener must use prosodic information, that is knowledge of the target talker's voice characteristics, and focus on that talker's voice. Erroneous percept of the target speech is considered to be caused by a combination of energetic and informational masking (Brungart et al., 2001). These complex sources of error notwithstanding, it is the auditory system's ability to discriminate residue pitch, including voice pitch, down to less than a tenth of a semitone (Micheyl et al., 2012) that gives the listener the main tool for understanding the target speech. And, as expected, the more dissimilar the voice fundamental frequency (f_0) and the idiosyncratic prosodic characteristics of the interfering talker from those of the target speaker, the more intelligible the target speech becomes (Brungart and Simpson, 2002). The number of interfering talkers is also important: the interference effect monotonically increases from 1 to 2 talkers but then it starts to decrease up to 16 interfering talkers (Rosen et al., 2013), because the larger the number of interfering talkers, the less intelligible their speech becomes. (By simple calculation it can be predicted that the output of 1,000 simultaneously but independently talking persons would be like a speech-spectrum noise between about 100 and 4,000 Hz; the auditory system is likely to treat interference by such a large number of talkers as if it were the random noise mentioned in the previous paragraph.)

But what does understanding speech in these cocktail party situations tell us about the auditory system's role in the perception of the dynamic output generated by the DRM or a DRM-like optimal production system? The answer is simple: just as the production of vowels and stop consonants was deduced and modeled from resonance properties of the vocal tract tube, the auditory system's decoding of the resonance-based speech signal also starts with the physical realities of the ear. Because the production system encodes the spectral envelope of the tube's resonances as a filtered response to the periodic excitation waveform originating in the vibration of the vocal folds, these filtered pulses excite frequency-specific parts of the basilar membrane which will generate appropriately scaled spectral excitation patterns that are essentially identical to the profile of the filtered tube resonances. When, instead of just one, there are two talkers each of whom is generating a particular resonance pattern that evolves dynamically over time, the acoustics of the two will mix and create a compound. The listener's ear will receive the compounded resonance patterns but, thanks to the auditory system's ability to perceive the vibrations of the two talkers vocal folds as two different pitches, it can separate the compound resonance patterns by associating one to the first, and the other to the second talker's voice pitch. Because the dynamic changes of the resonances along their respective formant trajectories proceed independently, they may often cross each other and generate perceptual errors – hence intelligibility of speech with interference by even one extraneous talker is reduced. Nevertheless, the listener can still follow the formant transitions, i.e., the dynamics of the talker's production system. Since, as suggested in the foregoing chapters, the speech production system can be considered optimal, by inference the speech perception system should be, too.

To conclude, this section has cited several results pointing to a view of the auditory system as particularly well suited for the perception of speech. But an apparatus optimally tuned for the reception of speech must also serve as a control mechanism supervising the production of speech – control as a sensorimotor feedback for the talker, and control by way of a behavioral reaction as a listener-originated feedback to the talker. Such a loop, in fact, forms a communication system fully consistent with the theory as first proposed by Claude Shannon (Shannon, 1948).

9 Epistemological considerations

In this book we undertook a deductive, iterative, and modeling approach to arrive at a novel understanding of speech. On the basis of its physical properties alone, a simple uniform acoustic tube was seen to become an *optimal* acoustic communication device. This transformation has occurred as a result of the tube acquiring modes and patterns of deformation over the course of a step-by-step iterative process self-guided by the sole criterion of always obtaining maximum acoustic variation using minimum deformation. The end result of this operation was a modeled system that, despite the fact that it came to existence without explicit reference to either the characteristics or the activity of the human system, was shown to be very close to the human speech production apparatus as we know it.

The following sections will summarize essential aspects of the preceding chapters, as regards the methodological triplet of deduction, iteration, and modeling, but will also show that the three form a logical chain coalescing into a single integrated entity that possesses specific relations between the articulatory, acoustic, phonetic, and phonological domains. The triplet represents an original approach that views speech as a fundamentally dynamic process. By way of this process (1) deformation gestures are mapped to acoustics with *maximum* efficiency dividing the acoustic tube into distinctive regions, (2) a model extracts acoustic trajectory primitives from the distinctive regions and structures the acoustic space, and (3) a coding scheme based on these primitives leads to an *acoustic phonology* that demonstrates a surprisingly close correspondence of this deduced phonology and the systems of oral vowels and plosive consonants used in present-day human communication. In other words, oral vowels and plosive consonants are intrinsic acoustic properties of a tube (Figure 9.1).

Because these comparisons between the human and the deduced optimal system have indicated that the two are remarkably similar, it appears logical to conclude – and this is the most important result – that the *human system* is likewise *acoustically optimal*, at least as far as oral vowels and plosive consonants are concerned. Furthermore, the deductive approach we used to build our modeled system has been throughout explanatory, it has always identified the logical foundation of each component of the system, always explained the results obtained, and always offered predictions for aspects of the model not yet observed.

Thus, our speech system today is acoustically quasi-optimal. Could this be sheer coincidence? Or could it be an ecological consequence of the communication process for which speech may have been used, that is, was the evolution of the speech system driven by communication needs? The answer to such questions should be important for gauging both the impact and the validity of our deduc-

DOI 10.1515/9781501502019-010

tions and predictions, although discussion of the evolutionary aspect will be deferred to the next chapter. One major advantage of our approach is that it uncovers cases that push the exploitation of sensitivity functions to its limits (i.e., from the control of small changes to dealing with large variations) and allows the ensuing hypotheses to be tested. The acoustic optimality observed for the production of oral vowels and plosive consonants provides an explanation for some characteristics of the speech production system and successfully predicts certain aspects of its behavior (such as the prediction of vocalic systems), although this does not imply that the optimality property can be extended to cover other sounds as well. However, after learning the places of articulation corresponding to oral vowels and plosive consonants and after these places are well integrated in the process, other sounds (fricatives, nasals, etc.) could also be produced with these places of articulation, provided that the acoustic contrast between the sounds that were kept will be sufficient, even if the acoustic optimality is not fully preserved. These other sounds would have to be viewed from the perspective of a dynamic process and they should exhibit sufficient acoustic contrast between one another.

In sum, we can reasonably assume that this optimality is the result of an evolutionary process driven by communication needs, which leads to the conclusion that oral vowels and plosive consonants are the *most efficient speech sounds*, i.e., optimal.

9.1 The integrated approach

Before discussing the epistemological relevance of the findings, components of our integrated process are listed and explained in the following paragraphs.

9.1.1 Deductive method

The use of deductive reasoning for the study of speech, starting with a given physical process, is not new. The best example is Fant (1960) who, with his theory of speech production, was among the first to establish a link between the shape of the vocal tract and resonant frequencies – representation of formant frequencies as a function of vocal tract shape can, indeed, provide the basis for explaining some, and predicting other aspects of vowel production. As another example, Liljencrants and Lindblom (1972) sought to predict vowel inventories from the largest possible spectral space obtained from models of the vocal tract and from acoustic distance measurements. But biological systems are inherently subject to changes resulting from adaptation to various environmental and genetic factors, so build-

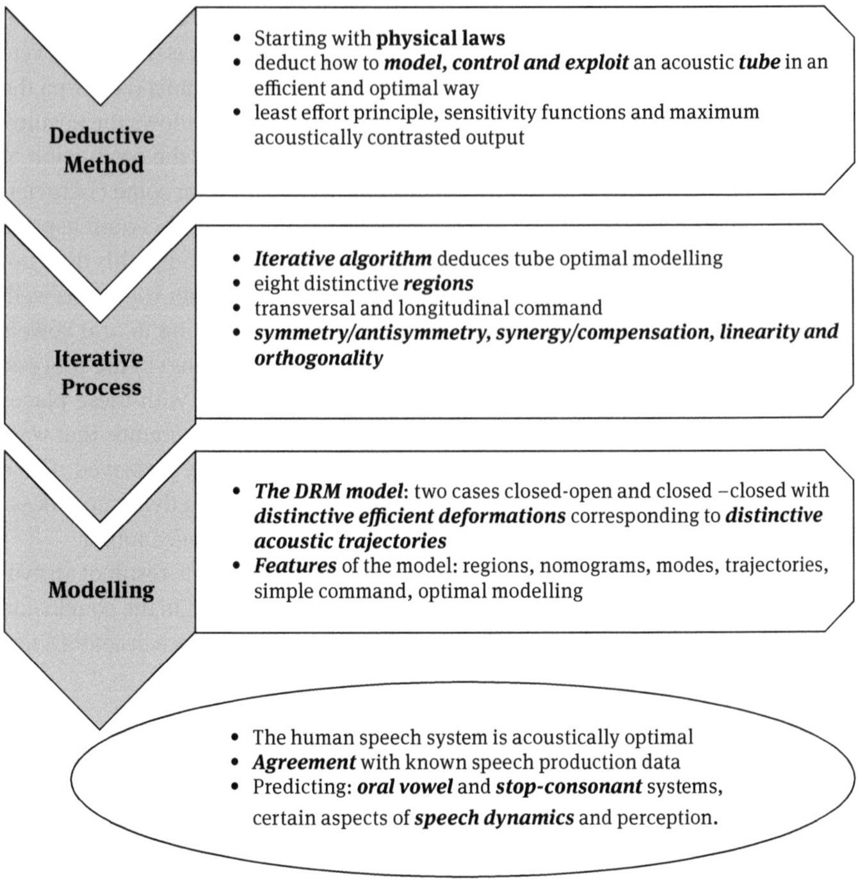

Fig. 9.1: The integrated deductive process/approach adopted in the book.

ing a model based on such systems could be inherently unstable. As described in Chapters 2 and 3, we start our deductive approach from permanent physical properties of a *uniform acoustic tube* without including any biologically imposed articulatory and perceptual constraints and, from this point of departure, we initiate a search for minimum deformations of the tube that are acoustically most effective.

9.1.2 Iterative process

The iterative algorithmic process described in Chapter 3 led step-by-step to minimal deformations of a uniform acoustic tube yielding maximal acoustic variation and halted when the acoustic change was no longer significant. The goal of these tube deformations was specified at the outset – e.g., increasing and decreasing F_1, then F_2, etc. – and each iterative step evolved toward this goal as the acoustic contrast between the initial and the final states of the tube gradually increased. However, as said throughout the previous chapters and explained in detail in Chapter 7, the process itself has never attempted to compare two static states; rather, the algorithm's goal was always dynamic – performing a gestural transition that resulted in a kinetic change of the tube's shape and produced a continuous pattern of formant change (temporarily continuous if each iterative step were converted to epochs on a time scale). Thus, all along this dynamic transition the goal remained the same and information on the dynamics of the transition (i.e., direction, rate, and total duration) was continuously available. The end result of the iterations is a tube deformation mapped into a vector in the acoustic space. The process aims at arriving at a maximum acoustic contrast between the tube configuration at the start and the one at the end of the iterations: maximum acoustic contrast observed today in humans defined by the space that the vowel triangle occupies. Inside this maximum space one can exploit sub-spaces that represent contrasts sufficient for achieving communication.

9.1.3 Modeling

Using the iterative process based on physical laws and modeling/control theory has led to an original model, the Distinctive Region Model (DRM). As described in detail in Chapter 4, the model is built on an acoustic tube with distinctive spatial regions. Exploiting the optimality of these regions, primitive gestures corresponding to trajectories in the acoustic domain are defined. The model permits the deduction of speech dynamics in *quantized domains*, both in tube area deformation space and the acoustic space: regions corresponding to maximally efficient segments and points of deformations are mapped into primitive trajectories structuring the acoustic space.

The model permits the production of acoustic transitions using *commands* that are simple, linear, and limited in number. The commands are region-specific and generate monotonic pseudo-orthogonal formant variations. The model ascertains and uses intrinsic acoustic features of the tube, such as anti-symmetry/symmetry, synergy, compensation, linearity, and orthogonality.

The importance of the modeling component in the integrated approach is that it allows asking questions that would test and expand our knowledge about speech: how do vowel and consonant systems develop, is phonology acoustic, and what determines and guides the complexity of these systems. The model can be used for *simply* generating synthesized sounds that are comparable, in a spectro-temporal domain, to those produced by the human articulatory system, and that are eminently suited for investigations of *timing, rate, and gesture synchrony.*

9.2 Findings

Our approach has uncovered several aspects and many important properties of speech by way of a deductive process based on physical laws, rather than being induced by observations. In spite of this, our approach has predicted essential aspects of speech as well as a number of findings that have been demonstrated by others as the fruit of a vast body of experimental work. We have also made it clear that the simulations obtained by our model will not replicate many details of the human speech communication system, since in its "vanilla" state it has no knowledge of a whole slate of parameters. Thus, in several cases (as shown, for example, throughout Chapter 7) perceptual data are used to fine-tune the model, and to adjust its behavior to make it reproduce data of real speech production and perception more faithfully. These preliminary remarks need to be kept in mind when, in the following subsections, we summarize findings with regard to dynamics, coding, timing, and complexity.

9.2.1 A dynamic approach to speech

The previous section represents an affirmation that our integrated approach is fundamentally theoretical, based on physical laws. Thus, it needs to be emphasized that although our system is compared to the human system producing real speech, it is not based on physical and physiological observations. Therefore, conclusions we draw from the similarities between the behavior of our system and data on speech production and perception do not imply that our system is a replica of the human one.

As explained in detail in Chapter 7, our iterative and modeling approach toward building an acoustic communication system has led to the discovery that, in order for the system to be efficient, the acoustic space must be structured in two ways. First, the structure contains a finite number of vocalic trajectories. Second, each of the consonant gestures constricting its proper distinctive region is mapped into a distinctive formant transition. Both types of trajectories and transitions (FTBs) are intrinsically dynamic and are generated by efficient gestural deformations of the acoustic tube. This dynamic approach has also led us to infer that the speech code consists of simple commands specifying the direction of the vocalic consonantal gestures.

If speech were nothing but a succession of events that occur at specific points of time, then precise measurements (difficult in a noisy environment) should be accessible at those specific instants. However, taking such measurements would require a solution to Zeno's Paradox (Huggett, 2010). In an effort to bypass trying to do the impossible, the dynamic approach avoids tasks that would attempt to find static targets and, instead, focuses on the transitions – whether in articulatory movements or in the acoustic spectrum. Taking this path offers a way to provide explanations for a host of not fully understood (and even controversial) phenomena, such as vowel reduction, hyper- and hypo-speech, and others. The rate of acoustic variation then becomes an important parameter. Whenever the rate is more or less constant along a formant transition, its value is accessible and measurable all along its course. Thus, characterizing a gesture by its direction (whether in the articulatory or in the acoustic space) and its velocity (of deformation in the articulatory space, or displacement in the acoustic space), views speech production as a genuinely dynamic task, in which objectives are not defined as targets in the acoustic or articulatory space but as time derivatives.

If speech representation is characterized by dynamic parameters, then it can be assumed that, at the production level, these parameters must be *invariant* to a certain degree (Verbrugge and Rackerd, 1986) and that the same parameters would also be used in the perceptual domain. In fact, as results of original experiments using formant trajectories with spectra outside the real F_1-F_2 vowel space demonstrated, even untrained listeners can identify and give appropriate labels to transitions along these trajectories (Chapter 7). Those results, incompatible with a static target approach, showed that such transitions can be categorized as vocalic trajectories according to the duration, the direction, and the rate of the transition. The same results also provide proof for the existence of a *dynamic invariance* and support the notion that listeners are able to interpret and use dynamic – velocity – parameters. In fact, consistent with these results, the existence of velocity (and acceleration) detectors in the auditory system has been demonstrated in psychoacoustic studies, as explained in Chapter 8 (Divenyi, 2005a; Crum and Hafter, 2008).

9.2.2 Coding and acoustic phonology

The primitives specified at the output of the model make it possible to establish a *dynamic code* that specifies which *gesture* (or ensemble of co-produced gestures) proceeds along what specific primitive *trajectory*, in what *direction*, and at what *rate*. This information is sufficient for creating a simple and efficient code that completely describes the speech sound to be generated. Thus, the epistemological significance of the coding operation is that it embodies a conversion: a much-needed link between phonetics and phonology. Details of this conversion are presented in Chapters 3, 4, and 5. Here we only want to summarize their relevance to coding.

One factor helping to establish a precise formulation of the acoustic-phonological relation is the dynamic quantal nature of the acoustic communication system. Quantization occurs on three main levels: (1) spatial quantization within the deformation gesture ranges inside the local areas related to production of vowels and consonants, (2) quantization in the acoustic domain along the seven trajectory primitives inside the structured vowel formant plane, (3) a time-domain quantization of the velocity and duration of formant transitions that make vowel–diphthong–consonant distinctions possible.

Places of articulation of both vowels and plosive consonants derive from the same model and are deduced from acoustic characteristics of the tube reflecting its inherent phonetic properties: oral vowels and plosive consonants, on the one hand, and phonetics and phonology, on the other, are intrinsically linked. Vowels are byproducts of vocalic trajectories and both vowels and plosive consonants can be obtained from specific gestural deformations of the tube that lead to specific formant trajectories in the acoustic plane. Simply stated, they are intrinsic acoustic properties of the tube. But since vowels are produced by a dynamic process, their widely used typological representations will not apply – they are irrelevant – and therefore should be reconsidered.

All this evokes the need to discuss the relationship between symbol and area function, viz. between phonology and phonetics. This relationship is generally considered weak and mediated on the cognitive level by some translation process, see, for example, O'Shaughnessy (1996, p. 1729) and Nearey (1997). In contrast, Fowler (1986) claimed that phonetic information is directly perceived from the acoustic signal.

The results and our analysis thereof are compatible with Fowler's claim because we deduced speech gesture primitives from inherent acoustic properties of the tube. By definition, gesture primitives are invariant and, in fact, a high degree of perceptual invariance of gestures is observed despite the relatively large variations in their production, such as inter-gestural timing, duration, and movement trajectory (see Chapter 7). These characteristics can vary within certain limits and, especially in the case of co-production, many (acoustic variants)-to-one (percept) can be obtained. These gestures are the ones that lead to a symbolic phonological code directly informed by phonetics. Then, as proposed by Fowler et al. (1980a), mediating translation going from symbol to articulation and to acoustics becomes unnecessary. Although Fowler's theory has not been without its critiques (e.g., Ohala, 1986; Lindblom, 1996), it is consistent with many aspects of our tube-plus-DRM system, in particular with the way two simultaneous gestures interact. This means that, in the case of the co-production of several gestures, a process analogous to deconvolution, one that separates and isolates the gestures should be present in the perceptual system (Fowler et al., 2016). With the speech signal being structured by the acoustic properties of the tube, this structuration can be perceived as such, that is, without any mediation.

If the speech communication system is explained by physical properties of the acoustic tube, then this system could be supported by the concept of a "single-strong theory" (around the acoustic tube, merging production and perceptual representation) instead of accepting the premises of a "double-weak theory" (weak on the levels of both production and perception, see Nearey, 1997). In our approach, speech gestures are specific deformations of the acoustic tube. When gestures are derived from data, for example, as Browman and Goldstein (1992) did, they remain articulatory. Then, instead of "articulatory phonology" suggested by these authors, we propose the concept of "acoustic phonology" be considered.

9.2.3 The importance of time

The formal (mathematical) significance of our dynamic approach is that practically all phenomena related to speech production and perception are examined, analyzed, and treated as temporal variations – as changes in time. The rates of formant transitions add important characteristics to vowels and consonants, and they should be more invariant (Verbrugge and Rackerd, 1986) than targets, maxima, minima, onsets or offsets of vowel formants (Hillenbrand and Nearey, 1999; Hillenbrand et al., 2001). Following such an approach, formant targets that are generally used to describe vowels would no longer be considered intrinsic but would become extrinsic. Instead, vowels and consonants would be characterized by dynamic parameters – their intrinsic property is the dynamics itself. Significant coarticulation effects, generally thought to be due to the attraction by static targets, would be reassigned to reflect transition rates. Normalization, generally performed in the frequency domain, would be transferred to the time domain (Verbrugge et al., 1976). The dynamic approach, in agreement with Fowler's (1980b) notion of intrinsic timing, thus assigns a primordial status to the temporal dimension. This approach assumes that dynamic time-domain parameters, such as transition rates, should be directly acquired by some learning process. Nevertheless, Lindblom et al. (2011) postulate that, in speech production, forces and directions can be obtained when specified targets are applied on a mechanism that translates the formant targets into equivalent motor commands (Saltzman and Munhall, 1989).

Direct reliance on dynamic parameters, rather than on static targets as Lindblom (2011) suggests, appears to be better adapted for explaining hyper- and hypo-speech. Such adaptive aspects could develop naturally within the frame of a sufficient acoustic space. As to the question how does one place a *sufficient* space inside the *maximum* space outlined by the vowel triangle, the answer is simple: around the neutral position! In fact, in Chapters 2 and 3 we show that the central, neutral configuration is acoustically the most efficient. This situation is exploited, for example, in vowel reduction (see Lindblom, 1963, who explicitly mentions centralization). Recognition of this relationship is an important result for speech dynamics.

9.2.4 Complexity

Because the single-strong code system is structured and quantized, the inherent complexity of speech communication could be explained and controlled. For example, properties of the DRM model offer a way to pilot vowel and consonant complexity: increasing the number of formants to be controlled, and consequently the number of gestures (by increasing the number of regions), results in increased complexity. In vowel production, adding the third formant F_3 as a coding element (recall that $F_3, F_4, \ldots$ always exist) increases complexity; this operation would add a complementary set of vowels and thus result in a set with twice the number of vowels compared to the one without F_3. The length of the regions to be controlled becomes finer and finer thus the articulatory system would become more and more accurate.

9.3 Conclusions

Figure 9.2 summarizes the epistemological significance of our deductive and iterative approach. At the central level, the abstract representation is syllabic. At that level this representation is invariant: planned coarticulation is not necessary. Although the model knows nothing about speech, deformations in the DRM are, in fact, speech gestures. Conversely, at the peripheral level the realization of these gestures is highly variable (one-to-many): they vary as a function of the context, of the synchrony/asynchrony of co-production, as well as of the individual (speaking rate, style, etc.). The listener's auditory system, however, is able to peel off all variability from the speech heard, files it as important to characterize the speaker but irrelevant to the message, and recovers the invariant dynamic information – performing an operation equivalent to deconvolution. Variability should be regarded as a necessary albeit secondary component of the communication between speaker and listener but may underlie the diachronic process of phonological sound changes. Such a direct link between the central and peripheral levels suggests that no mediating translation between the two is necessary.

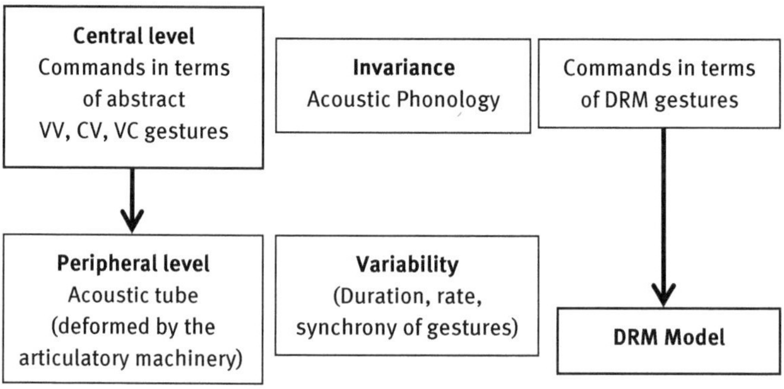

Fig. 9.2: Schematic representation of the theoretical connection between speech production stages. On the left: human speech production system; on the right: corresponding DRM modeling. There is a direct link between the central level and the periphery (no translation). Variability is observed at the peripheral level: the characteristics of the deformation gestures (size, movement, direction, and rate) may exhibit wide ranges of variation.

In conclusion, the epistemological significance of this book is that it endeavors to present a global, entirely dynamic, coherent, account of speech communication (oral vowels and consonant plosives) via a deductive approach based on the physics of an acoustic tube.

10 Conclusions and perspectives

This chapter summarizes the main message of the preceding chapters. It also presents major concepts and issues that are the outflow of our deductive-iterative modeling approach: optimality and the speech production and modeling process, speech as an intrinsically dynamic phenomenon, acoustic phonology, time versus frequency dimension, etc. It also suggests subjects for further investigation, such as the production of nasal consonants/fricatives, or the concept of evolution, and proposes possible applications in speech analysis, synthesis, and recognition.

10.1 Summary of findings and conclusions

10.1.1 Deductive-iterative modeling process

On the basis of physical laws, an iterative process deforming an acoustic tube leads to distinctive regions that form an optimal speech production model and build an acoustic communication system. The process is intrinsically dynamic. The model obtained can account for phenomena of speech communication: it can predict the main places of articulation for vowels and plosive consonants; it can predict gestural primitives and corresponding acoustic trajectory primitives leading to those speech sounds.

10.1.2 Efficiency and optimality

We have built an efficient and optimal model for the production of oral vowels and plosive consonants and found its results comparable to those obtained for human speech production. Therefore, we can conclude that the human production system approaches optimality (maximum possible efficiency) based on acoustic properties of a tube. However, human adaptation to his environment entails the acquisition and development of tools that make it possible to perform, in general, a large number of widely diverse tasks. For this purpose, humans have at their disposal a set of tools of general utility – in fact, having a tool specially designed for each task would result in creating a tool set of exceedingly high complexity. With a minimum number of tools, humans can perform a maximum number of tasks, certainly those necessary for their survival. If such a system is optimal, then its optimality would have to be global – a property rather difficult to formalize. In the case of speech, a robot capable of speaking would possess tools such as a jaw, a tongue, lips, an oral cavity coupled with nasal cavities, a pulse train generator,

DOI 10.1515/9781501502019-011

etc. However, a human being must also use the same tools for eating, drinking, laughing, and so on. Moreover, even if the robot happens to achieve optimality for the entire system of tasks, each task viewed in isolation may not be optimally performed. While we cannot tell if the human organ used for speaking is or is not optimal for all the other tasks it is involved in, the previous chapters have shown that, as far as the production of vowels and plosive consonants is concerned, there is ample evidence showing that the performance of the vocal apparatus is optimal or at least quasi-optimal. Nevertheless, these reservations notwithstanding, the deductive approach adopted in the book is able to keep all its explicative and predictive power.

10.1.3 Acoustic phonology

The deductively obtained DRM structures the acoustic tube into regions that correspond to standard places of articulation both for vowels and plosive consonants. These places have been derived using the same theoretical approach: deduction based on the acoustic characteristics of a tube reflecting its inherent phonetic properties. Thus, both oral vowels and plosive consonants can be obtained from specific gestural deformations of the tube that lead to specific formant trajectories in the acoustic domain. Consequently, oral vowels and plosive consonants actually represent intrinsic properties of an acoustic tube.

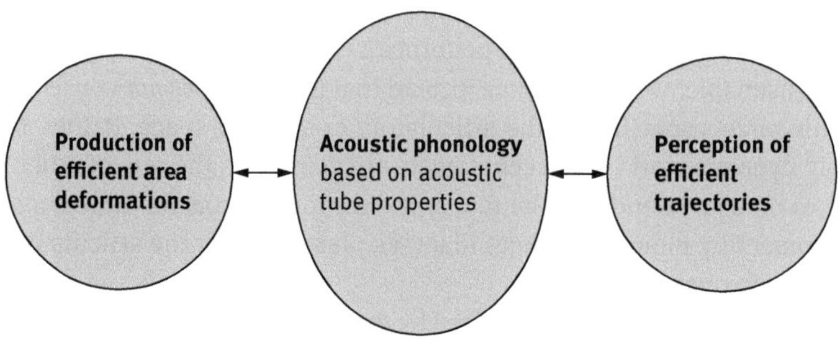

Fig. 10.1: Acoustic phonology is the core of the relation with production and perception.

The good correspondence between predicted and observed vocalic systems, and between predicted and observed plosive consonant places of articulation in speech production, leads us to take a hard look at the relations between symbol and area-function and, more importantly, between phonology and phonetics. Our approach derives deformation gestures of the tube from inherent acoustic properties of the tube. These gestures are directly linked to the phonetic code: there is no mediating translation from symbol to acoustics. Gestures can be considered to be symbolic primitives of the system, while vowels and consonants are products of the gestures. In other words, the theoretical framework for the production of oral vowels and plosive consonants, together with a direct link between phonetics and phonology leads to an acoustic phonology – i.e., a link between physical and linguistic structures in speech. Figure 10.1 summarizes the relationship between production and perception connected by acoustic phonology.

10.1.4 Dynamic process

If speech were nothing but a succession of events that occur at specific points of time – such as spectral extremes or onsets/offsets of transitions – then, in order to accurately characterize the events, precise measurements would have to be accessible at these specific instants. Such measurements become difficult or outright impossible in a noisy or reverberating environment. In the static approach, identification occurs only after the transition is completed, thus requiring a backward-integrative procedure to occur – a procedure especially difficult to complete under unfavorable acoustic conditions.

To go one step further, the phenomenon of vowel reduction assumes that, due to the high degree of inertia of the articulators, the articulatory mechanism is often unable to reach the intended targets. If reaching these targets were an absolute requirement, as many static theories imply, it would mean that the speech production mechanism is ill-adapted to perform an everyday task. Since this conclusion is unreasonable, we may assume instead that the task does *not* consist of reaching static targets specified in the articulatory or acoustic space. Rather, in the genuinely dynamic world of speech, targets are defined as the rate of articulatory gestures or, equivalently, as the rate and direction of acoustic transitions, that is, the moment-by-moment changes that take place in either the articulatory or in the formant frequency space.

If speech representation is characterized by dynamic parameters, then it can be assumed that, in the production domain, these parameters must be *invariant* at least to a certain degree and that the same invariant parameters are recovered in the perceptual domain. The dynamic approach we propose is able to naturally account for variability in a wide range of phenomena in addition to vowel reduction, such as the production of hyper- and hypo-speech, or the perception of male, female, and child speech – phenomena in each of which variability would have to be separately explained by different means.

In a dynamic approach, identification may occur already at the beginning of (and all along) the transition, using knowledge of the point of departure, trajectory direction, and transition velocity in the acoustic space. Although the point of departure can be acoustically known, to explain the perception of vowel trajectories outside the vowel triangle, it seems that they should also be phonologically specified. As shown in the experiment presented in Subsection 7.2.1.3, a given pair of F_1-F_2 formant frequencies can be perceived as /a/, /u/, or /o/ when reached by way of a transition originating at different departure points outside the vowel triangle. Clearly, the formant values of these points of departure do not correspond to any of those generally accepted as belonging to vowels, yet they are still perceived as vowels.

Recall that our dynamic process leads to vocalic primitive trajectories: vowels are byproducts of the trajectories. In such a dynamic approach, normalization should be performed in the time domain. Also, the static characteristics generally looked upon as intrinsic become extrinsic in this view, while dynamic characteristics are held as intrinsic. In general, the importance of the time domain in speech production and perception cannot be sufficiently emphasized.

10.1.5 Explanatory process

Being intrinsically deductive, our general approach is also explanatory. The acoustic properties of the tube can explain the main places of articulation for vowel and consonant production. Increasing the number of formants that are taken into account will make the production process more complex, while decreasing the number of formants will make it less so. The dynamic nature of the production system is inherently tied to the iterative process of arriving at a maximum acoustic effect at the cost of minimum effort. This approach can also predict new phenomena.

10.2 Perspectives

10.2.1 Extension of the model to speech sounds other than oral vowels and plosive consonants

As we have shown in the previous chapters, our tube-plus-DRM (model and control) system has allowed us to uncover the existence of deformations of regions leading to acoustically efficient trajectories; these regions possess a strong and direct link to the primary places of articulation of oral vowels and plosive consonants – speech sounds that are the most common in languages all over the world (see Maddieson, 1984). This strong acoustic link must have been recognized and taken advantage of in the evolutionary process, and, because they are used the most frequently, they could have provided a testing ground for the production of sounds not used previously. This means that, in principle, this new production could not be acoustically optimal. But these new sounds must be sufficiently different and the hypothesis can be proposed that the larger the acoustic difference between new sounds, and the larger it is between the new and the old sounds, the more frequently they are likely to be used.

This point was discussed in Subsection 5.5. The main places of articulation for consonant production are obtained from a closed-open tube. But the vowel [u], for example, is generated with a closed-(quasi-) closed tube. This means that the consonantal places of articulation in [uCu] are expected to change, or else they would not be able to keep the same formant transition characteristics as those obtained with the closed-open tube. In real speech, however, this is not what happens. What actually takes place is that the *optimal* places of articulation in the closed-open tube will dominate any other place. In particular, for the production of [uCu], the [u] vowel has a significant effect on the consonant transition, that is the pseudo-orthogonality of the formant transition behavior is not preserved. Regardless, the consonants in [uCu] are perceived clearly, suggesting that they are acoustically sufficiently distinct.

In view of the points raised above, the tube-plus-DRM system can be extended to the production of other types of speech sounds as well:

- Nasal vowels should be obtained from corresponding oral vowels with the velum opening added, but also placing them in a dynamic framework and verifying their acoustic distinctiveness.
- The principal places of articulation, and distinctive regions identified by the DRM for plosive consonants correspond also to those used for the production of other types of sounds, such as fricatives, affricates, nasals, and liquids (Mrayati et al., 1988). Again, these sounds will have to be viewed in a dynamic context and one would have to make sure that they exhibit sufficient distinctiveness, both within and across types.

In our approach, sound production is based on the dynamic anti-symmetrical deformation of an acoustic uniform tube of fixed length. However, it would be of interest to extend it to other types of deformations as well, such as increasing the length of the tube (both at the lip end and the pharynx end), in order to produce a rounding gesture; shifting the axis of the tube by disrupting its symmetry, in order to produce advanced tongue root (ATR, Carré et al., 2000); dividing one crucial distinctive region, R_6, into two parts with a movement of the tongue tip, in order to produce a synergetic effect that leads to a change in the F_4 formant (indispensable for retroflex plosives), etc.

10.2.2 Co-production-deconvolution of gestures

Gestural co-production means that two primitive gestures can be produced in parallel: for example, the speech signal corresponding to a CV syllable is the result of a command from the central level (abstract invariant consonant and vowel gestures) realized at the peripheral level (consonant and vowel gestural deformations in parallel) with varying duration, rate, and synchrony (Figure 9.1). The question is whether, during perceptual analysis of the speech signal, the syllable (produced by parallel gestures) is analyzed without attempting to decompose it into gestural components, or whether applying a deconvolution-like process can separate its two or more simultaneous components (see Fowler et al., 2016). Actually, we have unearthed a process analogous to deconvolution (see Figure 7.17, the perception of consonants around the neutral and its explanation in Chapter 7) and showed that equalizing, "neutralizing," a VCV utterance by shifting the vowels on either side of the consonant to neutral simplifies the processing of the transitions to and from the plosive: they will be identified by the listener based simply on transition directions (Carré et al., 2002). Chapters 3, 4, and 5 have demonstrated that assigning the uniform cross-sectional neutral vowel as the initial state for determining the maximum acoustic space and discovering the primitive trajectories leads to an acoustically optimal model and control of the tube. However, the just-cited VCV results suggest the hypothesis that the uniform tube shape of the acoustically efficient neutral has special characteristics that could be, or already may be, exploited in consonant production. Could such centralization also play a role in vowel production (vowel reduction)? Although the vowel [a] is not neutral (but still acoustically efficient – Figure 3.15) and its shape in the vocal tract is not uniform, it is the origin of primitive vocalic trajectories and, as such, it may have a specific role. As is widely observed, this vowel is the first one pronounced by the infant (de Boysson-Bardies, 1996) who, at birth, has a small pharynx cavity (Kent and Murray, 1982). To answer this question, extensive further research would be necessary but we believe that, once its importance is recognized, this research will eventually be carried out.

10.2.3 Transition slopes and time normalization

The large variability of static vowel targets is well known. Would relying on dynamic parameters of vowels bring us closer to invariance? At this point, the answer to this question is uncertain. Obviously, studies will be necessary to assess whether dynamic parameters display an invariance across male, female and child speech, across speech with various degrees of vowel reduction, across hyper- and hypo-speech, and across fast and normal speech.

As to normalization – a procedure performed in the frequency domain under a static approach – the dynamic approach we have taken leads to normalization in the time domain. While, to our knowledge, the efficiency of the two procedures has not been compared, it would be extremely informative if this were done in future experimental investigations.

Standard techniques used to recognize and track formant frequencies are notoriously imprecise and ill-suited for the analysis of speech under high-uncertainty conditions, such as high fundamental frequency or low SNR, and they measure spectral variations as their consequence rather than as their goal. Instead, one would need to develop analysis tools the primary objective of which is to register dynamic shifts of energy and of spectral configurations, in the time domain that also includes phase information (i.e., temporal fine structure, see Moore, 2008). Analysis of phase variation could become a tool for measuring the dynamics (i.e., the rate and direction) of transitions, continuing the work started by Chistovich et al. (1982), who used a simple auditory model to detect spectral transition bypassing formant tracking. Adopting a dynamic approach would compel speech research to reconsider analysis techniques, much in a way contemporary auditory research has.

10.2.4 From iterative to evolutionary process?

The deductive-iterative modeling process (Chapters 3 and 4) used for developing an efficient acoustic production system represents a step-by-step evolution toward an optimal state and use of the tube. As such, it shares many of its characteristics with a Lamarckian (Lamarck, 1968) evolutionary process (Figure 3.2) in which each iteration inherits properties of the previous one and thus can be regarded as an evolutionary step (Carré, Lindblom et al., 1995). From the first step on, and during the whole iterative process, the following changes take place:

(a) the distinctive region structure of the tube with distinctive, efficient deformations leads to the formulation of an optimal model: the DRM;
(b) at the same time, vocalic and consonant trajectory primitives naturally arise.

At the end of the deductive-iterative modeling process, the maximum of the acoustic space is obtained, the amplitudes of the deformation gestures and the acoustic primitive trajectories of the model also reach their respective maxima, efficient acoustic trajectories arise in the acoustic space, and a cavity analogous to that of the pharynx appears. The system that emerges can be considered optimal for the production of vowels and plosive consonants. Since the tube used to build this system and the one for human speech production are essentially identical, the human system can also be considered optimal. In fact, oral vowels and plosive consonants are intrinsic consequences of the acoustics of the tube.

Could our iterative algorithm be regarded as a model of the phylogenetic emergence of the human vocal tract? While it would be tempting to give it an evolutionary interpretation, several caveats appear to speak against it. First, our results have been obtained with a tube having an initially uniform shape – as shown in Subsection 3.2.4, the distinctive regions indispensable for speech cannot be obtained starting with a non-uniform tube (which was most probably the shape of the vocal tract *before* the descent of the larynx, see e.g., Lieberman et al., 1972). For the vocal tract to become such a tube, it would have had to be shaped with a 90° bend with a pharyngeal cavity, already at the beginning. Second, our distinctive regions were obtained iteratively starting from a uniform tube one formant at a time. Third, while this iterative process became optimal naturally for reasons inherent in our initial tube, any replica of such a process starting from a tube with an arbitrary shape would need a built-in evaluator that would determine if each evolutionary step had moved the system closer to optimality. To our knowledge, there are no existing theories of the evolution of speech that could overcome these three caveats.

Evolution of a tube like ours into the current human vocal tract would not be more likely either by postulating that optimal acoustic properties of the human system for the production of speech sounds (primitive gestural deformations of the tube shape with corresponding efficient acoustic trajectories that exploited properties of distinctive regions) could also have been discovered step-by-step during communication between speakers and listeners, as a result of some stochastic learning process (see for example Egan, 1957; Lenneberg, 1967; MacNeilage, 2008; Kirby and Sonderegger, 2013). In order for such learning to have taken place, the articulatory physiology and behavior of the vocal tract, coordinated by specialized processes of the brain, must have been well adapted for verbal communication from the outset. Computer models of various evolutionary scenarios have been offered by a number of authors (e.g., Steel, 1999; Oudeyer, 2005; de Boer, 2015; Moulin-Frier et al., 2015) – for an epistemological analysis see Studdert-Kennedy (1998). However, theories based on learning-driven processes are inductive rather than deductive and, to the extent that they aim at being explanatory, they cannot be fully so due to their inherent limitations. While the above cited and many other simulations have shown that a process-based stochastic or deterministic learning could evolve into a system of speech production and exploit it efficiently, such an approach, by its very nature, could not have knowledge of the *limits* of its own efficiency and thus *could not possess* the means for determining if the system is or is not acoustically optimal. Had we conducted a computer simulation experiment with our iterative system – an exercise that would have been easy to perform – it would have reached the same results as those presented in Chapters 3, 4, and 5, and thus that exercise would have been tautological.

Our objective was to push to the extreme an iterative approach anchored in a deductive framework, since it naturally lent itself to finding an optimal acoustic communication device in the most direct way. The approach we took was explanatory: it was based solely on physical criteria that have led to an acoustically optimal system. From that we observed that the human system behaves in a way similar to the one we have deduced and derived, and from this similarity we concluded that the human system must also be acoustically optimal or quasi-optimal. Yet, although all this can explain the characteristics of our device, it fails to provide a rationale that would account for the actual process of human evolution of speech, nor, as a matter of fact, for the diachronic sound changes proposed by phonologists. In contrast, when a tube is physiologically adapted for verbal communication, it is reasonable to suppose that there exists a uniform cross-sectional neutral position. Were this the case, the striving for optimality as this system is being exploited could have been boosted by the tube if it periodically returned to its neutral initial shape and compared to this ur-shape any newly acquired one. The possibility for such look-back-to-origin periods may not be unrealistic given the fact that periodically recurrent brief epochs of neutral shape have been observed in MRI recordings of fluent speech (Ramanarayanan et al., 2013). With a uniform tube serving as an omnipresent referent, one could thus imagine an evolutionary process of communication based on acoustic contrasts, first on those represented by F_1 changes, and then evolving in complexity as contrasts created by combined changes in F_1 and F_2, etc. are added, noting that all changes are also marked by the appearance of corresponding distinctive regions. That future evolutionary research might uncover findings in harmony with the above general scheme still remains a possibility.

10.2.5 Practical applications

The concept, characteristics, and ease of control of the simplified DRM (Section 4.5.1) makes it an appealing model for several applications in domains such as speech synthesis, analysis and recognition, inversion (speech signal to area function transform), and area shape constraints (such as speech when using a bite-block).

10.2.5.1 Speech Synthesis

Speech synthesis by the DRM can be performed in the time domain or in the frequency domain. In the *time domain*, the Kelly–Lochbaum model of wave propagation (Kelly and Lochbaum, 1962) can be used after uniformly discretizing all tubes, with excitation achieved by a two-mass model. The simplified DRM is able to provide efficient control for this model. In the *frequency domain*, the DRM can be coupled with a formant synthesizer and used as a tool for investigating various aspects of speech production. There are methods for calculating the transfer function for this composite DRM system, that is, the output pressure over the input volume velocity, and for obtaining the resonant frequencies (Badin and Fant, 1984; Mrayati, 1976). Both approaches can be used to synthesize vowels, semivowels, VV transitions, and VCV syllables; DRM model parameters can be verified and adjusted through perception tests. Beyond laboratory experiments, the DRM can also yield high-quality synthesis for private enterprise (e.g., Hill et al., 1995).

10.2.5.2 Speech Analysis and Recognition

The dynamic approach has also taken hold in the area of automatic speech recognition by introducing the use of dynamic cepstral coefficients (e.g., Furui, 1986a; see also Dusan, 2007). It is well known that dynamic parameters improve the recognition score. One method that can exploit the DRM concept and model is the Analysis-by-Synthesis Approach to Vocal Tract Modeling for Robust Speech Recognition (Al Bawab et al., 2008). Now, close to the direct perception point of view, is it possible to develop a speech recognition system that is able to deal with speech gestures (derived from DRM), formant frequencies, or maxima of spectral energy? Standard techniques generally developed in the frequency domain used for the recognition and tracking of formant frequencies are notoriously imprecise and ill-suited for the analysis of speech under high-uncertainty conditions, such as high fundamental frequency or low SNR. In addition, the DRM could inspire development of novel time-frequency analysis techniques, possibly using cochleagram-type displays (Lyon, 1983).

10.2.5.3 Inversion

The inversion problem, that is, determining the shape of the vocal tract from knowledge of acoustic parameters such as formant frequencies and bandwidths, has received quite a bit of attention over several decades. Unfortunately, there is no unique solution to the problem because there exist a large number of vocal tract configurations corresponding to a specific set of acoustic parameters. To find plausible solutions, various constraints need to be used and researchers have been trying to uncover methods for estimating the vocal tract shape from speech signals. Mermelstein (1967), Schroeder (1967), and Wakita (1979), just to cite a few among a large number of researchers, have proposed various approximate or partial solutions to this problem; generally though this problem was treated in the static situation. Looking for a dynamic solution, our work (Lammert et al., 2008) has relied on a codebook assembled from a spondee corpus resynthesized using the Haskins Laboratories Task Dynamic Application (TADA) synthesizer and a cochleagram analysis of the same corpus to extract a set of eight gestures.

The simplified DRM characterized only by a few parameters appears to be a suitable model for solving the inversion problem (see Soquet et al., 1990; Mrayati and Carré, 1992). In fact, our model has the potential of going much further: it could constitute an engine for a full-blown acoustic-to-articulatory inverse transform. The sensitivity functions of the iterative algorithm can be used to deform the tube in order to find a shape having any intended F-pattern that can be dynamic. The tube can be described in terms of a succession of elementary sections or in terms of the simplified DRM model that includes its own intrinsic constraints, such as synergy, reduced number of control parameters, etc. Places of articulation can be recovered from formant frequency changes. Such a technique was used by Story for "tuning" vocal tract area functions (Story, 2006).

10.2.5.4 Acoustic space and tube shape under specific constraints

Articulatory compensatory behavior in speech production can be investigated using the algorithm or the model. Bite-bloc speech is one example (Gay et al., 1981) showing that human beings are capable of modifying their manner of articulation in the presence of an obstruction in order to produce the same type of speech sounds that is used in real production. The speaker compensates for labial, dental, and jaw movements needed for some vowels – movements of the R_8 and partially of the R_7 and R_6 regions – by changes in the corresponding back regions, exploiting the compensatory feature between regions in the model (just as the ventriloquist does). In such cases, the maximum acoustic space can be obtained using the iterative algorithm: the values of the area range and the deformability coefficient are changed along the tube, according to the bite-block constraints.

The algorithm can also be used under specific constraints, for example as a tool for the simulation of glossectomy, in order to find the largest acoustic contrasts possible under specific anatomical constraints, and also to test prostheses (see, e.g., Carré, 1994) adapted to provide maximal acoustic space and maximal acoustic variations.

Bibliography

Abel SM. Discrimination of temporal gaps. J. Acoust. Soc. Am. 1972,52,519–524.

Ainsworth WA, and Carré R. Perception of second formant transitions in synthesized vowel sounds. Speech Comm. 1997,21,273–282.

Ainsworth WA. Discrimination of the shape of two-formant transitions before, between and after vowels. Acta Acustica united with Acustica 1999,85(1),121–127.

Al Bawab Z, Raj B, and Stern RM. Proceedings of 2008 IEEE International Conference on Acoustics, Speech and Signal Processing, 2008,4185–4188.

Al Dakkak O, Mrayati M, and Carré R. Transitions formantiques correspondant à des constrictions réalisées dans la partie arrière du conduit vocal [Formant transitions corresponding to constrictions executed in the rear part of the vocal tract]. Linguistica Communicatio 1994,6,59–63.

Arai T, and Greenberg S. The temporal properties of spoken Japanese are similar to those of English. Proc. Eurospeech '97, 1997,1011–1014.

Atal BS, and Hanauer SL. Speech Analysis and Synthesis by Linear Prediction of the Speech Wave. J. Acoust. Soc. Am. 1971,50,637–655.

Atal BS, Chang JJ, Mathews MV, and Tukey JW. Inversion of articulatory-to-acoustic transformation in the vocal tract by a computer-sorting technique. J. Acoust. Soc. Am. 1978,63,1535–1555.

Badin P, and Fant G. Notes on the vocal tract computations. KTH, STL-QPSR 1984,2–3,53–107.

Bagley WC. The apperception of the spoken sentence: A study in the psychology of language. Amer. J. Psychol. 1900,12,80–130.

Bakalla MH. A Chapter from the History of Arabic Linguistics: Ibn Jinnī: an early Arab Muslim phonetician. An interpretive study of his life and contribution to linguistics: Chapter from the history of Arabic linguistics. London: European Language Publisher, 1982.

Ballas JA. Common factors in the identification of an assortment of brief everyday sounds. J. Exp. Psychol. HPP 1993,19(2),250–267. 10.1037/0096-1523.19.2.250.

Barlow HB. Single units and sensation: A neuron doctrine for perceptual psychology? Perception 1972,1,371–394.

Benguerel A, and Cowan H. Coarticulation of upper lip protrusion in French. Phonetica 1974,30,41–55.

Blumstein SE, and Stevens KN. Acoustic invariance in speech production: Evidence from measurements of the spectral characteristics of stop consonants. J. Acoust. Soc. Am. 1979,66,1001–1017.

Bourdeau M, Dagenais L, and Santerre L. A quantitative framework for phonology. J. Appl. Ling. 1990,87–88,121–150.

Bregman AS. Auditory scene analysis: The perceptual organization of sound. Cambridge, MA: Bradford Books (MIT Press), 1990.

Broad DJ, and Clermont F. Linear scaling of vowel-formant ensembles (VFEs) in consonantal contexts. Speech Comm. 2002,37,175–195.

Bronkhorst AW. The Cocktail Party phenomenon: A review of research on speech intelligibility in multiple-talker conditions. Acta Acustica united with Acustica 2000,86(1),117–128.

Browder FE. Remarks on the direct method of the calculus of variations. Archives Mech. Anal. 1965,20,251–258.

DOI 10.1515/9781501502019-012

Browman CP, and Goldstein L. Towards an articulatory phonology. In: Ewan C, and Anderson J (Eds.), Phonology yearbook (Vol. 3). Cambridge: Cambridge University Press, 1986,219–252.

Browman CP, and Goldstein L. Articulatory phonology: An overview. Phonetica 1992,49,155–180.

Brown JC. Computer identification of musical instruments using pattern recognition with cepstral coefficients as features. J. Acoust. Soc. Am. 1999,105(3),1933–1941. doi: http://dx.doi.org/10.1121/1.426728.

Brownlee SA. The role of sentence stress in vowel reduction and formant undershoot: A study of lab speech and informal spontaneous speech. Unpublished Ph.D. Thesis, University of Texas, Austin, TX, 1996.

Brungart DS, Simpson BD, Ericson MA, and Scott KR. Informational and energetic masking effects in the perception of multiple simultaneous talkers. J. Acoust. Soc. Am. 2001,110(5 Pt 1),2527–2538.

Brungart DS, and Simpson BD. The effects of spatial separation in distance on the informational and energetic masking of a nearby speech signal. J. Acoust. Soc. Am. 2002,112(2),664–676. doi: http://dx.doi.org/10.1121/1.1490592.

Butcher A, and Weiher E. An electropalatographic investigation of coarticulation in VCV sequences. J. Phon. 1976,4,59–74.

Buus S. Release from masking caused by envelope fluctuations. J. Acoust. Soc. Am. 1985,78,1958–1965.

Byrd D, and Saltzman E. Intragestural dynamics of multiple prosodic boundaries. J. Phon. 1998,26(2),173–199. http://dx.doi.org/10.1006/jpho.1998.0071.

Carré R, and Mrayati M. Articulatory-acoustic-phonetic relations and modeling, regions and modes. In: Marchal A, and Hardcastle WJ (Eds.), Speech Production and Speech Modelling. Dordrecht: Kluwer Academic Publishers, 1990,211–240.

Carré R, and Mrayati M. Vowel-vowel trajectories and region modeling. J. Phon. 1991,19,433–443.

Carré R. A simulation tool for the study of the vowels after glossectomy, Abstracts Amer. Sp. Hear. Assoc.. New Orleans: American Speech and Hearing Association, 1994,86.

Carré R. About the fourth angle of the "vowel triangle". J. Acoust. Soc. Am. 1995,97,S3420.

Carré R, Bourdeau M, and Tubach JP. Vowel-vowel production: the distinctive region model (DRM) and vocalic harmony. Phonetica 1995,52,205–214.

Carré R, and Chennoukh S. Vowel-consonant-vowel modeling by superposition of consonant closure on vowel-to-vowel gesture. J. Phon. 1995,23,231–241.

Carré R, Lindblom B, and MacNeilage P. Rôle de l'acoustique dans l'évolution du conduit vocal humain [Acoustic factor in the evolution of the human vocal tract]. Comptes Rendus Acad. Sci., Paris 1995,30(IIb),471–476.

Carré R, and Mrayati M. Vowel transitions, vowel systems, and the Distinctive Region Model. In: Sorin C, Mariani J, Méloni H, and Schoetgen J (Eds.), Levels in Speech Communication: Relations and Interactions. Amsterdam: Elsevier, 1995,73–89.

Carré R, and Mody M. Prediction of vowel and consonant place of articulation. Proc. of the 3rd Meeting of the ACL Special Interest Group in Computational Phonology, SIGPHON 97, 1997,26–32.

Carré R. Perception of coproduced speech gestures. Proc. of the 14th Int. Cong. of Phonetic Sciences, 1999,643–646.

Carré R, and Divenyi P. Modeling and perception of "gesture reduction". Phonetica 2000,57,152–169.
Carré R, Sprenger-Charolles L, Messaoud-Galusi S, and Serniclaes W. On auditory-phonetic short-term memory. Proc. of the Int. Cong. of Speech and Language Processing (ICSLP2000) 2000,3,937–940.
Carré R, Ainsworth WA, Jospa P, Maeda S, and Pasdeloup V. Perception of vowel-to-vowel transitions with different formant trajectories. Phonetica 2001,58,163–178.
Carré R, Liénard JS, Marsico E, and Serniclaes W. On the role of the "schwa" in the perception of plosive consonants. Proc. of the 16th Int. Conf. on Speech and Language Processing, 2002,1681–1684.
Carré R. From acoustic tube to speech production. Speech Comm. 2004,42,227–240.
Carré R, Pellegrino F, and Divenyi P. Speech dynamics: epistemological aspects. Proc. 16th Intern. Congr. Phon. Sci., 2007,569–572.
Carré R. Production and perception of V_1V_2 described in terms of formant transition rates, Proc. of the Acoust. Soc. Amer. Meeting. Paris, 2008.
Carré R. Dynamic properties of an acoustic tube: Prediction of vowel systems. Speech Comm. 2009a,51,26–41.
Carré R. Signal dynamics in the production and perception of vowels. In: Pellegrino F, Marsico E, Chitoran I, and Coupé C (Eds.), Approaches to Phonological Complexity. Berlin: Mouton de Gruyter, 2009b,59–81.
Carré R. Coarticulation en production de parole: aspects acoustiques. In: Embarki M, and Dodane C (Eds.), La Coarticulation: Des indices à la représentation [Coarticulation: From signs of evidence to representation]. Paris: L'Harmattan, collection Langue & Parole, 2011,63–74.
Catford JC. A practical introduction to phonetics. Oxford: Clarendon Press, 1988.
Charfuelan M, and Schröder M. Investigating the prosody and voice quality of social signals in scenario meetings. In: D'Mello S, Graesser A, Schuller B, and Martin J-C (Eds.), Affective computing and intelligent interaction. Springer, 2011,46–56.
Cherry C. On human communication. Cambridge, MA: MIT Press, 1966.
Cherry EC. Some Experiments on the Recognition of Speech, with One and with Two Ears. J. Acoust. Soc. Am. 1953,25(5),975–979. doi: http://dx.doi.org/10.1121/1.1907229.
Cherry EC, and Taylor WK. Some further experiments upon the recognition of speech, with one and with two ears. J. Acoust. Soc. Am. 1954,26(4),554–559.
Chiba T, and Kajiyama M. The vowel. Its nature and structure. Tokyo: Tokyo-Kaiseikan Publishing Company, 1941.
Chistovich LA. Problems of speech perception. In: Hammerich LL, Jakobson R, and Zwirner E (Eds.), Form and Substance. Copenhagen: Akademisk Forlag, 1971,83–93.
Chistovich LA, Kozhevnikov VA, and Lesogor LV. Issledovanie perifericheskoi slukhovoi adaptatsii v psikhoakusticheskom eksperimente [Study of peripheral auditory adaptation in a psychoacoustic experiment]. Fiziol. Zh. SSSR Im. I M Sechenova 1972,58(10),1543–1547.
Chistovich LA, Fyodorova NA, Lissenko DM, and Zhukova MG. Auditory segmentation of acoustic flow and its possible role in speech processing. In: Fant G, and Tatham MAA (Eds). Auditory analysis and perception of speech. London: Academic, 1975,221–232.
Chistovich LA. Auditory processing of speech. Lang. Sp. 1980,23,67–72.
Chistovich LA, Lublinskaja VV, Malinikova TG, Ogorodnikova EA, Stoljarova EI, and Zhukov SJ. Temporal processing of peripheral auditory patterns of speech. In: Carlson R, and Grand-

ström B (Eds.), The representation of speech in the peripheral auditory system. Amsterdam: Elsevier Biomedical Press, 1982,165–180.

Chu S, Narayanan S, and Kuo CCJ. Environmental sound recognition with time-frequency audio features. IEEE Trans. ASSP 2009,17(6),1142–1158. 10.1109/TASL.2009.2017438.

Ciocca V, and Bregman AS. Perceived continuity of gliding and steady-state tones through interrupting noise. Perc. Psychoph. 1987,42(5),476–484. 10.3758/bf03209755.

Clements GN. Phonological primes: features or gestures. Phonetica 1992,49,181–193.

Clements GN. Lieu d'articulation des consonnes et des voyelles: une théorie unifiée [Places of articulation of consonants and vowels: a unified theory]. In: Laks B, and Rialland A (Eds.), L'Architecture des Représentations Phonologiques. Paris: CNRS Editions, 1993,147–171.

Clements GN. Feature economy as a phonological universal. Proc. 15th Intern. Congr. Phon. Sci., 2003,371–374.

Coker CH, Umeda N, and Browman CP. Automatic synthesis from ordinary English text. IEEE Transactions on Audio and Electroacoustics 1973, 21(3),293–298.

Cooper F, Liberman AM, and Borst JM. The inter-conversion of audible and visible patterns as a basis for research in the perception of speech. Proc. Natl. Acad. Sci. US 1951,37,318–325.

Cox DR. Algebra of probable inference. Baltimore: J.H. University Press, 1961.

Creelman CD. Human discrimination of auditory duration. J. Acoust. Soc. Am. 1962,34,582–593.

Crothers J. Typology and universals of vowel systems. In: Greenberg JH, Ferguson CA, and Moravcsik EA (Eds.), Universals of human language. Vol. 2: Phonology. Stanford: Stanford University Press, 1978,93–152.

Crum PA, and Hafter ER. Predicting the path of a changing sound: velocity tracking and auditory continuity. J. Acoust. Soc. Am. 2008,124(2),1116–1129. 10.1121/1.2945117.

Cutler A, Mehler J, Norris DG, and Segui J. The syllable's differing role in the segmentation of French and English. J. Mem. Lang. 1986,26,480–487.

Cutler A, and Norris DG. The role of strong syllables in segmentation for lexical access. J. Exp. Psychol. HPP 1988,14,113–121.

Cutler A, Mehler J, Norris D, and Segui J. Limits on bilingualism. Nature 1989,340(6230),229–230.

Cutler A, Melher J, Norris DG, and Segui J. The monolingual nature of speech segmentation by bilinguals. Cogn. Psychol. 1992,24,381–410.

D'Ausilio A, Pulvermüller F, Salmas P, Bufalari I, Begliomini C, and Fadiga L. The motor somatotopy of speech perception. Curr. Biol. 2009,19(5),381–385.

Dahl GE, Dong Y, Li D, and Acero A. Context-dependent pre-trained deep neural networks for large-vocabulary speech recognition. IEEE Trans. ASSP 2012,20(1),30–42. 10.1109/TASL.2011.2134090.

Daniloff R, and Hammarberg R. On defining coarticulation. J. Phon. 1973,1,239–248.

Dau T, Kollmeier B, and Kohlrausch A. Modeling auditory processing of amplitude modulation. I. Detection and masking with narrow-band carriers. J. Acoust. Soc. Am. 1997,102(5 Pt 1),2892–2905.

de Boer B. Self-organization in vowel systems. J. Phon. 2000,28,441–465.

de Boer B. Biology, culture, evolution and the cognitive nature of sound systems. J. Phon. 2015,53,79–87.

de Boysson-Bardies B. Comment la parole vient aux enfants. Paris: Odile Jacob, 1996.

de Cheveigné A. Separation of concurrent harmonic sounds: Fundamental frequency estimation and a time-domain concellation model of auditory processing. J. Acoust. Soc. Am. 1993,93,3271–3290.

de Cheveigné A, McAdams S, Laroche J, and Rosenberg M. Identification of concurrent harmonic and inharmonic vowels: A test of the theory of harmonic cancellation and enhancement. J. Acoust. Soc. Am. 1995,97(6),3736–3748. doi: http://dx.doi.org/10.1121/1.412389.

de Lamarck JM. Philosophie zoologique. Paris: 10/18, 1968.

de Saussure F. Cours de linguistique générale (1st edition 1916). Paris: Payot, 1995.

Dekerle M, Boulenger V, Hoen M, and Meunier F. Multi-talker background and semantic priming effect. Front. Hum. Neurosci., 2014,8. 10.3389/fnhum.2014.00878.

Delattre PC, Liberman AM, and Cooper FS. Acoustic loci and transitional cues for consonants. J. Acoust. Soc. Am. 1955,27,769–773.

Delhommeau K, Micheyl C, and Jouvent R. Generalization of frequency discrimination learning across frequencies and ears: Implications for underlying neural mechanisms in humans. J. Assoc. Res. Otolar. 2005,6(2),171–179. 10.1007/s10162-005-5055-4.

Demany L, and Semal C. Detection thresholds for sinusoidal frequency modulation. J. Acoust. Soc. Am. 1989,85(3),1295–1301.

Deng L. Dynamic speech models – Theory, algorithms, and applications: Morgan & Claypool Publishers, 2006.

Di Benedetto MG. Frequency and time variations of the first formant: Properties relevant to the perception of vowel height. J. Acoust. Soc. Am. 1989a,86,67–77.

Di Benedetto MG. Vowel representation: Some observations on temporal and spectral properties of the first formant frequency. J. Acoust. Soc. Am. 1989b,86,55–66.

Diehl RL, Lindblom B, and Creeger CP. Increasing realism of auditory representations yields further insights into vowel phonetics. Proc. of the 15th Intern. Congr. Phon. Sci., 2003,1381–1384.

Diehl RL, Lotto AJ, and Holt LL. Speech perception. Ann. Rev. Psychol. 2004,55,149–179. 10.1146/annurev.psych.55.090902.142028.

Divenyi PL, and Hirsh IJ. Identification of temporal order in three-tone sequences. J. Acoust. Soc. Am. 1974,56,146–151.

Divenyi PL, and Danner WF. Discrimination of time intervals marked by brief acoustic pulses of various intensities and spectra. Perc. Psychoph. 1977,21,125–142.

Divenyi PL, and Hirsh IJ. Some figural properties of auditory patterns. J. Acoust. Soc. Am. 1978,64,1369–1385.

Divenyi PL, and Sachs RM. Discrimination of time intervals bounded by tone bursts. Perc. Psychoph. 1978,24,429–436.

Divenyi PL, and Oliver SK. Resolution of steady-state sounds in simulated auditory space. J. Acoust. Soc. Am. 1989,85,2046–2056.

Divenyi PL, Lindblom B, and Carré R. The role of transition velocity in the perception of V1V2 complexes. Proc. 13th Intern. Congr. Phon. Sci., 1995,258–261.

Divenyi PL, and Haupt KM. Audiological correlates of speech understanding deficits in elderly listeners with mild-to-moderate hearing loss. III. Factor representation. Ear Hear. 1997a,18(3),189–201.

Divenyi PL, and Haupt KM. Audiological correlates of speech understanding deficits in elderly listeners with mild-to-moderate hearing loss. II. Correlation analysis. Ear Hear. 1997b,18(2),100–113.

Divenyi PL, and Haupt KM. Audiological correlates of speech understanding deficits in elderly listeners with mild-to-moderate hearing loss. I. Age and laterality effects. Ear Hear. 1997c,18(1),42–61.

Divenyi PL, and Carré R. The effect of transition velocity and transition duration on vowel reduction in V_1V_2 complexes. Proc. Joint Meet. Intern. Congr. Acoust. – Acoust. Soc. Amer., 1998,2957–2958.

Divenyi PL, and Brandmeyer A. The "cocktail-party effect" and prosodic rhythm: Discrimination of the temporal structure of speech-like sequences in temporal interference. In: Solé MJ, Recasens D, and Romero J (Eds.), Proc. 15th Intern.Congr. Phon. Sci. Barcelona, Spain, 2003,2777–2780.

Divenyi PL. The times of Ira Hirsh: Multiple ranges of auditory temporal perception. Semin. Hear. 2004,25(3),229–239.

Divenyi PL. Frequency change velocity detector: A bird or a red herring? In Pressnitzer D, Cheveigné A, and McAdams S (Eds.), Auditory Signal Processing: Physiology, Psychology and Models. New York: Springer-Verlag, 2005a,176–184.

Divenyi PL. Masking the feature-information in multi-stream speech-analogue displays. In: Divenyi PL (Ed.), Speech Separation by Humans and Machines. New York: Kluwer Academic Publishers, 2005b,269–281.

Divenyi PL (Ed.). Speech Separation by Humans and Machines. New York, NY: Kluwer Academic Publishers, 2005C.

Divenyi PL. Perception of complete and incomplete formant transitions in vowels. J. Acoust. Soc. Am. 2009,126(3),1427–1439. doi: http://dx.doi.org/10.1121/1.3167482.

Divenyi P. Decreased ability in the segregation of dynamically changing vowel-analog streams: a factor in the age-related cocktail-party deficit? Front. Neurosci. 2014,8(144). 10.3389/fnins.2014.00144.

Djarangar DI. Description phonologique et grammaticale du bédjonde, parler sara de Bédiondo/Tchad. Unpublished Doctoral dissertation in Linguistic Science [Thèse de doctorat en Sciences du Langage], Université Grenoble III, Grenoble, 1989.

Dooley GJ, and Moore BC. Duration discrimination of steady and gliding tones: a new method for estimating sensitivity to rate of change. J. Acoust. Soc. Am. 1988,84(4),1332–1337.

Dorman MF, Cutting, and Raphael LJ. Perception of temporal order on vowel sequences with and without formant transitions. J. Exp. Psychol. 1975,2,121–129.

Dorman MF, Studdert-Kennedy M, and Raphael LJ. Stop-consonant recognition: Release bursts and formant transitions as functionally equivalent, context-dependent cues. Perc. Psychoph. 1977,22,109–122.

Drullman R, Festen JM, and Plomp R. Effect of reducing slow temporal modulations on speech reception. J. Acoust. Soc. Am. 1994a,95(5 Pt 1),2670–2680.

Drullman R, Festen JM, and Plomp R. Effect of temporal envelope smearing on speech reception. J. Acoust. Soc. Am. 1994b,95(2),1053–1064.

Duez D. On spontaneous French speech aspects of the reduction and contextual assimilation of voiced stops. J. Phon. 1995,23,407–427.

Durlach NI, Mason CR, Kidd, G Jr, Arbogast TL, Colburn HS, and Shinn-Cunningham BG. Note on informational masking. J. Acoust. Soc. Am. 2003,113(6),2984–2987.

Dusan S. On the relevance of some spectral and temporal patterns for vowel classification. Speech Communication 2007,49,71–82.

Eddins DA, Hall JW. III, and Grose JH. The detection of temporal gaps as a function of frequency region and absolute noise bandwidth. J. Acoust. Soc. Am. 1992,91,1069–1077.

Egan JP. Monitoring task in speech communication. J. Acoust. Soc. Am. 1957,29(4),482–489. doi: http://dx.doi.org/10.1121/1.1908936.

Ehrenfest P. On adiabatic changes of a system in connection with the quantum theory. Proc. Amsterdam Acad. 1916,19,576–597.
Elliott L, Hammer M, Scholl M, Carrell TD, and Wasowicz J. Discrimination of rising and falling simulated single-formant frequency transitions: Practice and transition duration effects. J. Acoust. Soc. Am. 1989,86,945–953.
Elman JL, and McClelland JL. Exploiting lawful variability in the speech wave. In: Perkell JS, and Klatt DH (Eds.), Invariance and variability in speech processes. Hillsdale: Erlbaum, 1986,360–385.
Fant CGM. On the predictability of formant levels and spectrum envelopes from formant frequencies. In: Halle M, Lunt H, and MacLean H (Eds.), For Roman Jakobson. The Hague: Mouton, 1956,109–120.
Fant CGM. Acoustic theory of speech production. The Hague: Mouton, 1960.
Fant CGM, and Màrtony J. The Instrumentation for Parametric Synthesis (OVE II). Speech Trans. Labs. QPSR 1962,3(2),18–24.
Fant CGM. Auditory patterns of speech. In: Dunn WW (Ed.), Models for the perception of speech and visual form. Cambridge, MA: MIT Press, 1967,111–125.
Fant CGM. Acoustic tract wall effects, losses and resonance bandwidths. STL-QPSR 1972,2–3,28–52.
Fant CGM. Speech sounds and features. Cambridge, MA: MIT Press, 1973.
Fant CGM, and Pauli S. Spatial characteristics of vocal tract resonance modes. Proc. Speech Commun. Semin., 1974,121–132.
Fant CGM. Vocal tract area and length perturbations. STL-QPSR 1975a,4,1–14.
Fant CGM. Non-uniform vowel normalization. STL-QPSR 1975b,16,1–19.
Fant CGM. Key-Note Speech. U.S.–Japan joint seminar on dynamic aspects of speech production 1976,17(4),21–27.
Fant CGM. The relations between area functions and the acoustic signal. Phonetica 1980,37,55–86.
Farnetani E, Vagges K, and Magno-Caldognetto E. Coarticulation in Italian /VtV/ sequences: a palatographic study. Phonetica 1985,42(2–3),78–99.
Farnetani E, and Recasens D. Coarticulation models in recent speech production theories. In: Hardcastle WJ, and Hewlett N (Eds.), Coarticulation theory, data and techniques. Cambridge: Cambridge University Press, 1999,31–68.
Fastl H. Fluctuation strength and temporal masking patterns of amplitude-modulated broadband noise. Hear. Res. 1982,8(1),59–69. http://dx.doi.org/10.1016/0378-5955(82)90034-X.
Festen JM, and Plomp R. Effects of fluctuating noise and interfering speech on the speech-reception threshold for impaired and normal hearing. J. Acoust. Soc. Am. 1990,88,1725–1736.
Firth IA. Modal analysis of the vocal tract. STL-QPSR 1986,2–3,1–12.
Flanagan JL. Speech Analysis Synthesis and Perception. Berlin: Springer-Verlag, 1972.
Flanagan JL. Difference limen for the intensity of a vowel sound. J. Acoust. Soc. Am. 1955a,27(6),1223–1225. doi: http://dx.doi.org/10.1121/1.1908174.
Flanagan JL. A difference limen for vowel formant frequency. J. Acoust. Soc. Am. 1955b,27,613–617.
Flanagan JL. Band width and channel capacity necessary to transmit the formant information of speech. J. Acoust. Soc. Amer. 1956,28(4),592–596. doi: http://dx.doi.org/10.1121/1.1908412.

Flanagan JL. Note on the design of terminal-analog speech synthesizers. J. Acoust. Soc. Am. 1957,29307–310.

Flege JE. Effects of speaking rate on tongue position and velocity of movement in vowel production. J. Acoust. Soc. Am. 1988,84,901–916.

Fletcher H. Auditory patterns. Review of Modern Physics 1940,12,47–65.

Fogerty D, and Kewley-Port D. Perceptual contributions of the consonant-vowel boundary to sentence intelligibility. J. Acoust. Soc. Am. 2009,126(2),847–857. doi: http://dx.doi.org/10.1121/1.3159302.

Fogerty D, and Humes LE. Perceptual contributions to monosyllabic word intelligibility: Segmental, lexical, and noise replacement factors. J. Acoust. Soc. Am. 2010,128(5),3114–3125. doi: http://dx.doi.org/10.1121/1.3493439.

Fontanini A, and Katz DB. Behavioral states, network states, and sensory response variability. J. Neurophys. 2008,100(3),1160–1168. 10.1152/jn.90592.2008.

Fowler CA, Rubin P, Remez R, and Turvey MT. Implications for speech production of the general theory of action. In: Butterworth B (Ed.), Speech Production I: Speech and talk. London: Academic Press, 1980a,373–420.

Fowler CA. Coarticulation and theories of extrinsic timing. J. Phon. 1980b,8,113–133.

Fowler CA. An event approach to the study of speech perception from a direct-realist perspective. J. Phon. 1986,14,3–28.

Fowler CA, and Rosenblum LD. The perception of phonetic gestures. Haskins Lab. Sp. Rep., SR-99/100, 1989,102–117.

Fowler CA, Shankweiler D, and Studdert-Kennedy M. Perception of the speech code revisited: Speech is alphabetic after all. Psychol. Rev. 2016,123,125–150.

Freyman RL, Balakrishnan U, and Helfer KS. Effect of number of masking talkers and auditory priming on informational masking in speech recognition. J. Acoust. Soc. Am. 2004,115(5),2246–2256. doi: http://dx.doi.org/10.1121/1.1689343.

Frick RW. Communicating emotion: The role of prosodic features. Psychol. Bull. 1985,97(3),412–429. 10.1037/0033-2909.97.3.412.

Fujisaki H, and Kawashima T. A model of the mechanisms for speech perception. Quantitative analysis of categorical effects in discrimination. Ann. Rep. Eng. Res. Inst. 1971,30,59–68.

Fujisaki H, and Sekimoto S. Perception of time-varying resonance frequencies in speech and non-speech stimuli. In: Cohen A, and Nooteboom S (Eds.), Structure and Process in Speech Perception (Vol. 11). Berlin Heidelberg: Springer Verlag, 1975,269–282.

Fujisaki H. On the modes and mechanisms of speech perception – Analysis and interpretation of categorical effects in discrimination. In: Lindblom B, and Öhman S (Eds.), Frontiers of Speech Communication Research. London: Academic, 1979,177–189.

Furui S. On the role of spectral transition for speech perception. J. Acoust. Soc. Am. 1986a,80,1016–1025.

Furui S. Speaker-independant isolated word recognition using dynamic features of speech spectrum. IEEE Trans. ASSP 1986b,34,52–59.

Furukawa S, and Moore BCJ. Dependence of frequency modulation detection on frequency modulation coherence across carriers: Effects of modulation rate, harmonicity, and roving of the carrier frequencies. J. Acoust. Soc. Am. 1997,101(3),1632–1643. doi: http://dx.doi.org/10.1121/1.418147.

Gay T. Articulatory movements in VCV sequences. J. Acoust. Soc. Am. 1977,62,183–193.

Gay T. Effect of speaking rate on vowel formant movements. J. Acoust. Soc. Am. 1978,63,223–230.

Gay T, Lindblom B, and Lubker J. Production of bite-block vowels: acoustic equivalence by selective compensation. J. Acoust. Soc. Am. 1981,69,802–810.
Ghitza O. Processing of spoken CVCs in the auditory periphery. I. Psychophysics. J. Acoust. Soc. Am. 1993,94(5),2507–2516. doi: http://dx.doi.org/10.1121/1.407386.
Ghitza O, and Greenberg S. On the possible role of brain rhythms in speech perception: intelligibility of time-compressed speech with periodic and aperiodic insertions of silence. Phonetica 2009,66(1–2),113–126. 000208934 [pii] 10.1159/000208934.
Gibson JJ. Outline of a theory of direct visual perception. In: Royce JR, and Rozeboom WW (Eds.), The psychology of knowing. New York: Gordon & Breach, 1972.
Gick B, and Wilson I. Excrescent schwa and vowel laxing: Cross-linguistic responses to conflicting articulatory targets. In: Goldstein LM, Whalen DH, and Best CT (Eds.), Laboratory Phonology VIII. Berlin: Mouton de Gruyter, 2006,635–659.
Giraud A-L, and Poeppel D. Cortical oscillations and speech processing: emerging computational principles and operations. Nat. Neurosci. 2012,15(4),511–517.
Godfrey JJ, Syrdal-Lasky AK, Millay KK, and Knox CM. Performance of dyslexic children on speech perception tests. J. Exp. Child Psychol. 1981,32,401–424.
Goldstein JL. Auditory Nonlinearity. J. Acoust. Soc. Am. 1967,41(3),676–699. doi: http://dx.doi.org/10.1121/1.1910396.
Green DM. Profile analysis: Auditory intensity discrimination. Oxford: Oxford University Press, 1988.
Greenberg S. Speaking in shorthand: A syllable-centric perspective for understanding pronunciation variation. Speech Comm. 1999,29,159–176.
Grey J. Multidimensional scaling of musical timbres. J. Acoust. Soc. Am. 1976,61,1270–1277.
Griffiths TD, and Warren JD. What is an auditory object? Nat. Rev. Neurosci. 2004,5(11),887–892.
Grimault N, Micheyl C, Carlyon RP, Arthaud P, and Collet L. Influence of peripheral resolvability on the perceptual segregation of harmonic complex tones differing in fundamental frequency. J. Acoust. Soc. Am. 2000,108(1),263–271. doi: http://dx.doi.org/10.1121/1.429462.
Guenther FH. Speech sound acquisition, coarticulation, and rate effects in a neural network model of speech production. Psychol. Rev. 1995,102,594–621.
Gunnilstam O. The theory of local linearity. J. Phon. 1974,2,91–108.
Gygi B, and Shapiro V. Development of the Database for Environmental Sound Research and Application (DESRA): Design, functionality, and retrieval considerations. Proc. EURASIP Audio Sp. Mus. 2010,2010,654914.
Hanna TE. Effect of signal phase and masker separation on two-tone masking. J. Acoust. Soc. Am. 1986,80,473–478.
Harris KS, Hoffman HF, Liberman AM, Delattre PC, and Cooper FS. Effect of third-formant transitions on the perception of the voiced stop consonants. J. Acoust. Soc. Am. 1958,30,122–126.
Harshman R, Ladefoged P, and Goldstein L. Factor analysis of tongue shapes. J. Acoust. Soc. Am. 1977,62,693–707.
Heinz JM. Perturbation functions for the determination of vocal tract area functions from vocal tract eigenvalues. STL-QPSR 1967,1,1–14.
Hermansky H. Speech recognition from spectral dynamics. Sadhana 2011,36,729–744.
Hill D, Manzara L, and Taube-Schock CR. Real-time articulatory speech-synthesis-by-rules. Paper presented at the Proc. AVIOS 11–14 Sept. '95, San Jose, 1995.

Hillenbrand JM, Minifie FD, and Edwards TJ. Tempo of spectrum change as a cue in speech-sound discrimination by infants. J. Sp. Hear. Res. 1979,22(1),147–165.

Hillenbrand JM, Getty LA, Clark MJ, and Wheeler K. Acoustic characteristics of American English vowels. J. Acoust. Soc. Am. 1995,97,3099–3111.

Hillenbrand JM, and Nearey TM. Identification of resynthesized /hVd/ utterances: Effects of formant contour. J. Acoust. Soc. Am. 1999,105,3509–3523.

Hillenbrand JM, Clark MJ, and Nearey TM. Effects of consonant environment on vowel formant patterns. J. Acoust. Soc. Am. 2001,100,748–763.

Hirsh IJ. Auditory perception of temporal order. J. Acoust. Soc. Am. 1959,31,759–767.

Hirsh IJ. Temporal order and auditory perception. In: Moskowitz HR, Scharf B, and Stevens JC (Eds.), Sensation and measurement: Papers in honor of S. S. Stevens. Dordrecht, The Netherlands: D. Reidel, 1974,251–258.

Hockett C. Manuel of phonology, Publications in Anthropology and Linguistics (Vol. 11). Bloomington: Indiana University Press, 1955.

House AS, and Fairbanks G. The influence of consonant environment upon the secondary acoustical characteristics of vowels. J. Acoust. Soc. Am. 1953,25,105–113.

Houtgast T. Frequency selectivity in amplitude-modulation detection. J. Acoust. Soc. Am. 1989,85(4),1676–1680. doi: http://dx.doi.org/10.1121/1.397956.

Huang CB. Perception of first and second formant frequency trajectories in vowels. Proc. 11th Int. Congr. Phon. Sci., 1987,194–197.

Huggett N. Zeon's paradoxes. In: Zalta EN (Ed.), The Stanford Encyclopedia of Philosophy (Winter 2010 edition). Stanford, CA: Stanford University, 2010.

Iskarous KR. Dynamic acoustic-articulatory relations. Unpublished Ph.D., University of Illinois, Urbana-Champaign, 2001.

Jakobson R, Fant CGM, and Halle M. Preliminaries to speech analysis: The distinctive features and their correlates. Cambridge: The MIT Press, 1951.

Jakobson R, and Halle M. Fundamentals of language. The Hague: Mouton, 1956.

Jenkins JJ, Strange W, and Miranda S. Vowel identification in mixed-speaker silent-center syllables. J. Acoust. Soc. Am. 1994,95,1030–1694.

Johnson K. Speaker perception without speaker normalization. An exemplar model. In: Johnson K, and Mullennix JW (Eds.), Talker Variability in Speech Processing. New York: Academic Press, 1997,145–165.

Johnson-Davies D, and Patterson RD. Psychophysical tuning curves: Restricting the listening band to the signal region. J. Acoust. Soc. Am. 1979,65(3),765–770. doi: http://dx.doi.org/10.1121/1.382490.

Johnsrude IS, Mackey A, Hakyemez H, Alexander E, Trang HP, and Carlyon RP. Swinging at a Cocktail Party: Voice Familiarity Aids Speech Perception in the Presence of a Competing Voice. Psychol. Sci. 2013,24(10),1995–2004. 10.1177/0956797613482467.

Kay RH, and Matthews DR. On the existence in human auditory pathways of channels selectively tuned to the modulation present in frequency-modulated tones. J. Physiol. 1947,225,657–677.

Kelly J, and Lochbaum C. Speech synthesis. Proc. 4th Intern. Congr. Acoust., G 42, 1962,1–4.

Kent RD, and Moll KL. Vocal-tract characteristics of the stop cognates. J. Acoust. Soc. Am. 1969,46,1549–1555.

Kent RD, and Minifie FD. Coarticulation in recent speech production models. J. Phon. 977,5,115–133.

Kent RD, and Murray A. Acoustic features of infant vocalic utterances at 3, 6, and 9 months. J. Acoust. Soc. Am. 1982,72(2),353–364.
Kewley-Port D. Measurement of formant transitions in naturally produced stop consonant-vowel syllables. J. Acoust. Soc. Am. 1982,73,379–389.
Kewley-Port D, and Zheng Y. Auditory models of formant frequency discrimination for isolated vowels. J. Acoust. Soc. Am. 1998a,103(3),1654–1666.
Kewley-Port D, Burkle TZ, and Lee JH. Contribution of consonant versus vowel information to sentence intelligibility for young normal-hearing and elderly hearing-impaired listeners. J. Acoust. Soc. Am. 2007,122(4),2365–2375. 10.1121/1.2773986.
Kidd, G Jr, Mason CR, Deliwala PS, Woods WS, and Colburn HS. Reducing informational masking by sound segregation. J. Acoust. Soc. Am. 1994,95(6),3475–3480.
Kidd, G Jr, Mason CR, Best V, and Marrone N. Stimulus factors influencing spatial release from speech-on-speech masking. J. Acoust. Soc. Am. 2010,128(4),1965–1978. 10.1121/1.3478781.
Kirby J, and Sonderegger M. A model of population dynamics applied to phonetic change. In: Knauff M, Pauen M, Sebanz N, and Wachsmuth I (Eds.), 35th Annual Conference of the Cognitive Science Society. Austin, TX: Cognitive Science Society, 2013,776–781.
Kjems U, Boldt JB, Pedersen MS, Lunner T, and Wang D. Role of mask pattern in intelligibility of ideal binary-masked noisy speech. J. Acoust. Soc. Am. 2009,126(3),1415–1426. doi: http://dx.doi.org/10.1121/1.3179673.
Klatt DH. Speech perception: A model of acoustic-phonetic analysis and lexical access. J. Phon. 1979,7,279–312.
Kluender KR, and Jenison RL. Effects of glide slope, noise intensity, and noise duration on the extrapolation of FM glides through noise. Percept Psychophys 1992,51(3),231–238.
Kluender KR, Coady JA, and Kiefte M. Sensitivity to change in perception of speech. Speech Comm. 2003,41(1),59–69. http://dx.doi.org/10.1016/S0167-6393(02)00093-6.
Koening W, Dunn HK, and Lacey LY. The sound spectrograph. J. Acoust. Soc. Am. 1946,18,19–49.
Koffka K. Principles of Gestalt psychology. New York: Harcourt Brace Jovanovich, 1935.
Kozhevnikov VA, and Chistovich LA. Speech, articulation, and perception (JPRS-30543): NTIS, US Dept. of Commerce, 1965.
Krull D. Second formant locus pattern and consonnant-vowel coarticulation in spontaneous speech. Perilus, X, 87–108, 1989.
Kuehn DP, and Moll KL. A cineradiographic study of VC and CV articulatory velocities. J. Phon. 1976,4,303–320.
Kuhl PK, and Meltzoff AN. The intermodal representation of speech in infants. Infant Behav. Devel. 1984,7,361–381.
Kuroda T, Nakajima Y, and Eguchi S. Illusory continuity without sufficient sound energy to fill a temporal gap: Examples of crossing glide tones. J. Exp. Psychol. HPP 2012,38(5),1254–1267. 10.1037/a0026629.
Kwatny HG. Nonlinear Control and Analytical Mechanics. Boston: Birkhaeuser, 2000.
Labov W. Sociolinguistics Patterns. Philadelphia: University of Pennsylvania, 1972.
Ladefoged P, and Lindau M. Where does the vocal tract end ? UCLA WWP 1979,45,32–38.
Ladefoged P, and Maddieson I. The sounds of the world's languages. Oxford, UK: Blackwell Publishers, 1996.
Lammert A, Ellis DPW, and Divenyi P. Data-driven articulatory inversion incorporating articulator priors. Proc. 4th SAPA Workshop, Interspeech '08, 2008.

Lass R. Vowel system universals and typology: Prologue to theory, Phonology Yearbook (Vol. 1). Cambridge: University Press, 1984,75–111.
Lawrence W. The Synthesis of Speech from Signals which have a Low Information Rate. In: Jackson W (Ed.), Communication Theory. London: Butterworth, 1953,460–469.
Lehiste I, and Peterson GE. Vowel amplitude and phonemic stress in American English. J. Acoust. Soc. Am. 1959,31(4),428–435. doi: http://dx.doi.org/10.1121/1.1907729.
Lehiste I, and Peterson GE. Transitions, glides, and diphthongs. J. Acoust. Soc. Am. 1961,33,268–277.
Lemaitre G, Houix O, Misdariis N, and Susini P. Listener expertise and sound identification influence the categorization of environmental sounds. J. Exp. Psychol. Appl. 2010,16(1),16–32. 10.1037/a0018762.
Lenneberg EH. Biological foundations of language. New York: Wiley, 1967.
Liberman A, Cooper F, Shankweiler D, and Studdert-Kennedy M. Perception of the speech code. Psychol. Rev. 1967,74,431–461.
Liberman AM, Delattre PC, Cooper FS, and Gerstman LJ. The role of consonant vowel transitions in the perception of the stop and nasal consonants. Psychol. Monogr. 1954,68,1–13.
Liberman AM. Some results of research in speech perception. J. Acoust. Soc. Am. 1957,29,117–123.
Liberman AM, Harris KS, Hoffman HS, and Griffith BC. The discrimination of speech sounds within and accross phoneme boundaries. J. Exp. Psychol. 1957,54,358–368.
Liberman AM, and Mattingly IG. The motor theory of speech perception revisited. Cognition 1985,21,1–36.
Lieberman P, Crelin ES, and Klatt D. Ability and related anatomy of the newborn and adult human, Neanderthal man and chimpanzee. American Anthropologist 1972,74,287–307.
Liljencrants J, and Lindblom B. Numerical simulation of vowel quality systems: The role of perceptual contrast. Language 1972,48,839–862.
Lindblom B. Spectrographic study of vowel reduction. J. Acoust. Soc. Am. 1963,35,1773–1781.
Lindblom B, and Studdert-Kennedy M. On the role of formant transitions in vowel perception. J. Acoust. Soc. Am. 1967,42,830–843.
Lindblom B, and Sundberg J. Acoustic consequences of lip, tongue, jaw and larynx movement. J. Acoust. Soc. Am. 1971,50,1166–1179.
Lindblom B. On the origin and purpose of discreteness and invariance in sound patterns. In: Perkell J, and Klatt DH (Eds.), Invariance and invariability in speech processes. Hillsdale, New Jersey: Erlbaum, 1986a,493–523.
Lindblom B. Phonetic universal in vowel systems. In: Ohala JJ, and Jaeger JJ (Eds.), Experimental Phonology. Orlando: Academic Press, 1986b,13–43.
Lindblom B. Phonetic contents in phonology. Perilus, XI, 101–118, 1990a.
Lindblom B. Explaining phonetic variation: a sketch of the H and H theory. In: Marchal A, and Hardcastle WJ (Eds.), Speech Production and Speech Modelling, NATO ASI Series. Dordrecht: Kluwer Academic Publishers, 1990b,403–439.
Lindblom B. Phonetic content in phonology. Paper presented at the Phonologica 1988, Proc. of the 6th Int. Phonology Meeting, Krems, Austria, 1992.
Lindblom B, Guion S, Hura S, Moon S-J, and Willerman R. Is sound change adaptive? Revista di Linguistica 1995,7,5–37.
Lindblom B. Role of articulation in speech perception: Clues from production. J. Acoust. Soc. Am. 1996,99,1683–1692.

Lindblom B, Brownlee SA, and Lindgren R. Formant undershoot and speaking styles: an attempt to resolve some controversial issues. In: Simpson AP, and Pätzold M (Eds.), Sound Patterns of Connected Speech, Models and Explanation. Kiel: Institute of Phonetics and Digital Speech Processing, University of Kiel, 1996,119–130.

Lindblom B, Davis JH, Brownlee SA, Moon SJ, and Simpson Z. Energetics in phonetics: A preliminary look. Proc. of LP '98, Prague, 1999,401–415.

Lindblom B, Diehl R, Park S-H, and Salvi G. Sound systems are shaped by their users: The recombination of phonetic substance. In: Clements GN, and Ridouane R (Eds.), Where Do Phonological Features Come From?: Cognitive, physical and Developmental Bases of Distinctive Speech Categories. Amsterdam/Philadelphia: John Benjamins Publishing Company, 2011,67–98.

Lindgren R, and Lindblom B. Reduction of vowel chaos. TMH-QPSR 1996,2,1–4.

Lippmann RP. An introduction to computing with neural nets. IEEE Trans. ASSP 1987,4(2),4–22. 10.1109/MASSP.1987.1165576.

Lutfi RA. A model of auditory pattern analysis based on component-relative-entropy. J. Acoust. Soc. Am. 1993,94(2),748–758. doi: http://dx.doi.org/10.1121/1.408204.

Lyon RF. A computational model of binaural localization and separation. Proc. IEEE ICASSP, 1983,1148–1151.

Lyzenga J, and Carlyon RP. Center frequency modulation detection for harmonic complexes resembling vowel formants and its interference by off-frequency maskers. J. Acoust. Soc. Am. 1999,105(5),2792–2806.

Macmillan NA. Beyond categorical/continuous distinction: A psychophysical approach to processing modes. In: Harnad S (Ed.), Categorical perception. New York: Cambridge University Press, 1987,53–85.

Macmillan NA, and Creelman CD. Detection theory: A user's guide. Cambridge, U.K.: Cambridge University Press, 1991.

MacNeilage PF. The frame/content theory of evolution of speech production. Behav. Br. Sci. 1998,21,499–546.

MacNeilage PF. The origin of speech. Oxford, U.K.: Oxford University Press, 2008.

Maddieson I. Patterns of sounds. Cambridge: Cambridge University Press, 1984.

Maeda S. Un modèle articulatoire basé sur une étude acoustique. Bulletin de l'IPG 1979,8,35–55.

Maeda S. Compensatory articulation during speech; evidence from the analysis and synthesis of vocal-tract shapes using an articulatory model. In: Marchal A, and Hardcastle WJ (Eds.), Speech Production and Speech Modelling, NATO ASI Series. Dordrecht: Kluwer Academic Publishers, 1990,131–149.

Mann VA, and Liberman A. Some differences between phonetic and auditory modes of perception. Cognition 1983,14,211–235.

Mattingly IG. The global character of phonetic gesture. J. Phon. 1990,18,445–452.

McClelland JL, and Elman JL. The TRACE model of Speech Perception. Cogn. Psychol. 1986,18,1–86.

McKeon R. Aristotle's conception of language and the arts of language. Classic. Philol. 1946,41(4),193–206.

Mermelstein P. Determination of the vocal tract shape from measured formant frequencies. J. Acoust. Soc. Am. 1967,41,1283–1294.

Mermelstein P. Articulatory model for the study of speech production. J. Acoust. Soc. Am. 1973,53,1070–1082.

Micheyl C, Ryan CM, and Oxenham AJ. Further evidence that fundamental-frequency difference limens measure pitch discrimination. J. Acoust. Soc. Am. 2012,131(5),3989–4001. doi: http://dx.doi.org/10.1121/1.3699253.

Miller JL, and Liberman A. Some effects of later-occuring information on the perception of stop consonant and semivowel. Perc. Psychoph. 1979,25,457–465.

Miller JL. Phonetic perception: evidence for context-dependent and context-independent processing. J. Acoust. Soc. Am. 1981,69(3),822–831.

Mills JW. On the minimum audible angle. J. Acoust. Soc. Am. 1958,30,237–246.

Moore BCJ, and Glasberg BR. Suggested formulae for calculating auditory-filter bandwidths and excitation patterns. J. Acoust. Soc. Am. 1983,74(3),750–753. doi: http://dx.doi.org/10.1121/1.389861.

Moore BCJ, and Glasberg BR. Formulae describing frequency selectivity as a function of frequency and level, and their use in calculating excitation patterns. Hear. Res. 1987,28(2),209–225. http://dx.doi.org/10.1016/0378-5955(87)90050-5.

Moore BCJ, Glasberg BR, and Schooneveldt GP. Across-channel masking and comodulation masking release. J. Acoust. Soc. Am. 1990,87,1683–1694.

Moore BCJ, Sek A, and Shailer MJ. Modulation discrimination interference for narrow-band noise modulators. J. Acoust. Soc. Am. 1995,97(4),2493–2497.

Moore BCJ. The role of temporal fine structure processing in pitch perception, masking, and speech perception for normal-hearing and hearing-impaired people. J. Assoc. Res. Otolar. 2008,9(4),399–406. 10.1007/s10162-008-0143-x.

Moore BCJ. An introduction to the psychology of hearing (6th edn.). Leiden, the Netherlands: Brill, 2013.

Morrison GS, and Assmann PF (Eds.). Vowel inherent spectral change. Berlin: Springer, 2013.

Moulin-Frier C, Diard J, Schwartz JL, and Bessière P. COSMO ("Communicating about Objects using Sensory–Motor Operations"): A Bayesian modeling framework for studying speech communication and the emergence of phonological systems. J. Phon. 2015,53,6–41.

Mrayati M. Contribution aux etudes sur la production de la parole. Modeles electriques du conduit vocal avec pertes. Etude de leurs interactions. Relations entre disposition articulatoire et caracteristiques acoustiques. Grenoble: These de Doctorat-es-Sciences, 1976.

Mrayati M, and Carré R. Relations entre la forme du conduit vocal et les caractéristiques acoustiques des voyelles françaises. Etude des distributions spatiales [Relations between the form of the vocal tract and French vowels' acoustic characteristics. A study of spatial distributions]. Phonetica 1976,33,285–306.

Mrayati M, Guérin B, and Boë LJ. Etude de l'impédance d'entrée du conduit vocal, couplage source-conduit vocal [Study on the entrance impedance of the vocal tract, source-vocal tract coupling]. Acustica 1976,35,330.

Mrayati M, Carré R, and Guérin B. Distinctive regions and modes: A new theory of speech production. Speech Comm. 1988,7,257–286.

Mrayati M, Carré R, and Guérin B. Distinctive regions and modes: articulatory-acoustic-phonetic aspects. A reply to Boë and Perrier. Speech Comm. 1990,9,231–238.

Mrayati M, and Carré R. The acoustic-area function inversion problem and the distinctive region model. Paper presented at the Proc. of EUSIPCO-92, Bruxelles, 1992.

Narayanan S, Toutios A, Ramanarayanan V, Lammert A, Kim J, Lee S, Nayak K, Kim Y-C, Zhu Y, Goldstein L, Byrd D, Bresch E, Ghosh P, Katsamanis A, and Proctor M. Real-time magnetic resonance imaging and electromagnetic articulography database for speech production

research (TC). J. Acoust. Soc. Am. 2014,136(3),1307–1311. doi: http://dx.doi.org/10.1121/1.4890284.
Nearey TM, and Assmann P. Modeling the role of inherent spectral change in vowel identification. J. Acoust. Soc. Am. 1986,80,1297–1308.
Nearey TM. Static, dynamic, and relational properties in vowel perception. J. Acoust. Soc. Am. 1989,85,2088–2113.
Nearey TM. Speech perception as pattern recognition. J. Acoust. Soc. Am. 1997,101,3241–3254.
Nord L. Acoustic studies of vowel reduction in Swedish. STL-QPSR 1986,4,19–36.
O'Shaughnessy D. Critique: Speech perception: Acoustic or articulatory. J. Acoust. Soc. Am. 1996,99,1726–1729.
Oh EL, and Lutfi RA. Informational masking by everyday sounds. J. Acoust. Soc. Am. 1999,106(6),3521–3528.
Ohala J. Moderator's introduction to symposium on phonetic universals in phonological systems and their explanations. Proc. 9th Intern. Congr. Phon. Sci., III, 1979,181–185.
Ohala J. The emergent syllable. In: Davis BL, and Zajdo K (Eds.), Syllable Development: The Frame/Content Theory and beyond. Mahwah, N.J.: Lawrence Erlbaum Associates, 2008,179–186.
Ohala JJ. The listener as a source of sound change. In: Masek RA, Hendrick RA, and Miller MF (Eds.), Papers from the Parasession on Language and Behavior. Chicago: Chicago Ling. Soc, 1981,178–203.
Ohala JJ. Against the direct realist view of speech perception. J. Phon. 1986,14,75–82.
Ohala JJ. What's behind the smile? Behav. Br. Sci. 2010,33(06),456–457. doi: 10.1017/S0140525X10001585.
Öhman SEG. Coarticulation in VCV utterances: spectrographic measurements. J. Acoust. Soc. Am. 1966a,39,151–168.
Öhman SEG. Perception of segments of VCCV utterances. J. Acoust. Soc. Am. 1966b,40,979–988.
Oppenheim A. Speech Spectrograms Using The Fast Fourier Transform. IEEE Spectrum 1970,7(8),57–62.
Oudeyer P-Y. The Self-Organization of Speech Sounds. J. Theor. Biol. 2005,233,435–449.
Peterson GE, and Barney HL. Control methods used in the study of the vowels. J. Acoust. Soc. Am. 1952,24,175–184.
Plomp R. The ear as a frequency analyzer. J. Acoust. Soc. Am. 1964,36,1628–1636.
Plomp R. Timbre as a multidimensional attribute of complex tones. In: Plomp R, and Smoorenburg GF (Eds.), Frequency analysis and periodicity detection in hearing. Leiden: Sijthoff, 1970,393–414.
Plomp R. Perception of speech as a modulated signal. In: van der Broecke MPR, and Cohen A (Eds.), Proceedings of the Tenth International Congress of Phonetic Sciences. Dordrecht: Foris, 1983,29–40.
Poeppel D. The analysis of speech in different temporal integration windows: cerebral lateralization as 'asymmetric sampling in time'. Speech Comm. 2003,41(1),245–255. http://dx.doi.org/10.1016/S0167-6393(02)00107-3.
Pollack I, Rubenstein H, and Horowitz A. Communication of verbal modes of expression (Vol. 3). London: Kingston Press Services, 1960.
Pollack I. Detection of rate of change of auditory frequency. J. Exp. Psychol. 1968,77,535–541.
Pols LCW, and van Son RJJH. Acoustics and perception of dynamic vowel segments. Speech Comm. 1993,13,135–147.

Pulvermuller F, Huss M, Kherif F, Moscoso del Prado Martin F, Hauk O, and Shtyrov Y. Motor cortex maps articulatory features of speech sounds. Proc. Natl. Acad. Sci. USA 2006,103(20),7865–7870. 0509989103 [pii] 10.1073/pnas.0509989103.
Ralston JV, and Sawusch JR. Perception of sine wave analogs of stop consonant place information. J. Acoust. Soc. Am. 1984,76,S28.
Ramanarayanan V, Goldstein L, Byrd D, and Narayanan SS. An investigation of articulatory setting using real-time magnetic resonance imaging. J. Acoust. Soc. Am. 2013,134(1),510–519. doi: http://dx.doi.org/10.1121/1.4807639.
Rasmussen CE, and Ghahramani Z. Occam's razor. Adv. Neural Inform. Proc. Syst. 2001,13,294–300.
Recasens D. Vowel-to-vowel coarticulation in Catalan VCV sequences. J. Acoust. Soc. Am. 1984,76,1624–1635.
Repp BH. The role of psychophysics in understanding speech perception. In: Schouten MEH (Ed.), The psychophysics of speech perception. Dordrecht, the Netherlands: Nijhoff, 1987,3–27.
Roederer J. Introduction to the physics and psychophysics of music. New York: Springer, 1975.
Rosen S, Souza P, Ekelund C, and Majeed AA. Listening to speech in a background of other talkers: Effects of talker number and noise vocoding. J. Acoust. Soc. Am. 2013,133(4),2431–2443. doi: http://dx.doi.org/10.1121/1.4794379.
Roweis S. One microphone source separation. Neural Inform. Proc. Syst. 2000,13,793–799.
Sachs M, Young E, and Miller M. Encoding of speech features in the auditory nerve. In: Carlson R, Granström B (Eds.), The Representation of Speech in the Peripheral Auditory System. Amsterdam: Elsevier Biomedical, 1982,115–130.
Saltelli A. Sensitivity analysis for importance assessment. Risk Anal. 2002,22(3),1–12.
Saltzman EL. Task dynamic coordination of the speech articulators: a preliminary model. In: Heuer H, and Fromm C (Eds.), Generation and modulation of action patterns (Vol. 15). New York: Springer-Verlag, 1986,129–144.
Saltzman EL, and Munhall KG. A dynamic approach to gestural patterning in speech production. Ecol. Psychol. 1989,1,333–382.
Sankoff D and Sankoff G. Sample survey methods and computer-assisted analysis in the study of grammatical variation. In Darnell R (ed.), Canadian Languages in their Social Context. Edmonton, Alberta: Linguistic Research, 1973,7–64.
Scharf B. Partial masking. Acta Acustica united with Acustica 1964,14(1),16–23.
Scherer KR. Nonlinguistic vocal indicators of emotion and psychopathology. In: Izard CE (Ed.), Emotions in personality and psychopathology. New York: Plenum, 1979,495–529.
Schouten JF. The residue and the mechanism of hearing. Proc. Royal Acad. Sci. (Netherl.) 1940,43,356–365.
Schouten MEH. The case against a speech mode of perception. Acta Psychol. 1980,44,71–98.
Schouten MEH. Identification and discrimination of sweep tones. Perc. Psychoph. 1985,37(4),369–376. 10.3758/BF03211361.
Schouten MEH, and Pols LCW. Identification and discrimination of sweep formants. Perc. Psychoph. 1989,46,235–244.
Schouten MEH (Ed.). The psychophysics of speech perception. The Hague: Nijhoff, 1987.
Schroeder MR. Determination of the geometry of the human vocal tract by acoustic measurements. J. Acoust. Soc. Am. 1967,41,1002–1010.
Schroeder MR, and Strube J. Acoustic Measurements of Articulator Motions. Phonetica 1979,36(3),302–313.

Schroeder MR, and Strube HW. Flat-spectrum speech. J. Acoust. Soc. Am. 1986,79,1580–1583.
Schwartz J-L, Boë L-J, Vallée N, and Abry C. Major trends in vowel system inventories. J. Phon. 1997a,25,233–253.
Schwartz JL, Boë LJ, Vallée N, and Abry C. The dispersion-focalization theory of vowel systems. J. Phon. 1997b,25,255–286.
Shannon CE. A mathematical theory of communication. The Bell System Technical Journal 1948,27,379–423, 623–656.
Shinn-Cunningham BG, Ihlefeld A, Satyavarta, and Larson E. Bottom-up and top-down influences on spatial unmasking. Acta Acustica united with Acustica 2005,91(6),967–979.
Sluijter AMC, and van Heuven VJ. Spectral balance as an acoustic correlate of linguistic stress. J. Acoust. Soc. Am. 1996,100(4),2471–2485. doi: http://dx.doi.org/10.1121/1.417955.
Soklakov AN. Occam's razor as a formal basis for a physical theory. Foundations of Physics Letters 2002,15,107–135.
Soquet A, Saerens M, and Jospa P. Acoustic-articulatory inversion based on a neural controller of a vocal tract model. Proceedings of the ESCA Workshop on Speech Synthesis. ESCA, 1990,71–74.
Steel L. The spontaneous self-organization of an adaptive language. In: Muggleton S (Ed.), Machine Intelligence (Vol. 15). Oxford: Oxford Univiversity Press, 1999,205–224.
Stevens KN, and House AS. Development of a quantitative description of vowel articulation. J. Acoust. Soc. Am. 1955,27,484–493.
Stevens KN. The quantal nature of speech: evidence from articulatory-acoustic data. In: David EE, and Denes PB (Eds.), Human Communication: a unified view. New York: Mac Graw–Hill, 1972,51–66.
Stevens KN, and Blumstein SE. The search for invariant acoustic correlates of phonetic features. In: Eimas PD, and Miller JL (Eds.), Perspectives on the study of speech. Hillsdale: Erlbaum, 1981.
Stevens KN. Evidence for the role of acoustic boundaries in the perception of speech sounds. In: Fromkin VA (Ed.), Phonetic Linguistics. Essay in Honor of Peter Ladefoged. Orlando: Academic Press, 1985,243–255.
Stevens KN. On the quantal nature of speech. J. Phon. 1989,17,3–45.
Stevens KN. Acoustic phonetics. Cambridge, MA: MIT Press, 2000.
Stevens SS. A scale for the measurement of the psychological magnitude: loudness. Psychol. Rev. 1936,43(5),405–416.
Story B. Technique for "tuning" vocal tract area functions based on acoustic sensitivity functions. J. Acoust. Soc. Am. 2006,119,715–718.
Strange W, Verbrugge RR, Shankweiler DP, and Edman TR. Consonant environment specifies vowel identity. J. Acoust. Soc. Am. 1976,60,213–224.
Strange W, Jenkins JJ, and Johnson TL. Dynamic specification of coarticulated vowel. J. Acoust. Soc. Am. 1983,74,695–705.
Strange W. Dynamic specification of coarticulated vowels spoken in sentence context. J. Acoust. Soc. Am. 1989a,85,2135–2153.
Strange W. Evolving theories of vowel perception. J. Acoust. Soc. Am. 1989b,85,2081–2087.
Studdert-Kennedy M, Knight C, and Hurford JR. Introduction: new approaches to language evolution. In: Hurford JR, Studdert-Kennedy M, and Knight C (Eds.), Approaches to the Evolution of Language. Cambridge: Cambridge University Press, 1998,1–5.
Sundberg J, Johansson C, Wilbrand H, and Ytterbergh C. From sagittal distance to area. A study of transverse, vocal tract cross-sectional area. Phonetica 1987,44,76–90.

ten Bosch LFM. On the structure of vowel systems: aspects of an extended vowel model using effort andcontras. Unpublished doctoral dissertation, University of Amsterdam, Amsterdam, 1991.
ten Bosch LFM, Bonder LJ, and Pols LCW. Static and dynamic structure of vowel systems. Proc. 11th Intern. Congr. Phon. Sci., 1986,235–238.
Terhardt E. Pitch, consonance, and harmony. J. Acoust. Soc. Am. 1974,55,1061–1069.
Tran Thi AX, Nguyen VS, Castelli E, and Carré R. Production and perception of pseudo-V_1CV_2 outside the vowel triangle: Speech illusion effects. Paper presented at the Interspeech 2013, Lyon, 2013.
Trubetzkoy NS. Principles of phonology. Berkeley, CA: University of California Press, 1969.
Turk AE, and Shattuck-Hufnagel S. Word-boundary-related duration patterns in English. J. Phon. 2000,28(4),397–440. http://dx.doi.org/10.1006/jpho.2000.0123.
Ungeheuer G. Elemente einer akustischen Theory der Vokalarticulation [Elements of an acoustic theory of vowel articulation]. Berlin: Springer-Verlag, 1962.
Uzgalis W. John Locke. In: Zalta EN (Ed.), The Stanford Encyclopedia of Philosophy. Stanford, CA: Stanford University, 2016.
Valentine S, and Lentz JJ. Broadband auditory stream segregation by hearing-impaired and normal-hearing listeners. J. Speech Lang. Hear Res 2008,51(5),1341–1352.
van Noorden LPAS. Temporal coherence in the perception of tone sequences. Unpublished doctoral dissertation, Technische Hoogschool, Eindhoven (the Netherlands), 1975.
van Noorden LPAS. Minimum differences of level and frequency for perceptual fission of tone sequences ABAB. J. Acoust. Soc. Am. 1977,61,1041–1045.
van Son RJJH, and Pols LCW. Formant movements of Dutch vowels in a text, read at normal and fast rate. J. Acoust. Soc. Am. 1992,92,121–127.
van Son RJJH, and Pols LCW. Vowel identification as influenced by vowel duration and formant track shape. Proc. Eurospeech '93, 1993,285–288.
van Son RJJH, and van Santen JPH. Duration and spectral balance of intervocalic consonants: A case for efficient communication. Speech Comm. 2005,47(1–2),100–123. http://dx.doi.org/10.1016/j.specom.2005.06.005.
van Wieringen A, and Pols LC. Transition rate as a cue in the perception of one-formant speech-like synthetic stimuli. Proc. 12th Intern. Congr. Phon. Sci., 1991,446–449.
Verbrugge RR, Strange W, Shankweiler DP, and Edman TR. What information enables a listener to map a talker's vowel space. J. Acoust. Soc. Am. 1976,60,198–212.
Verbrugge RR, and Rakerd B. Talker-independent information for vowel identity. Haskins Lab. Sp. Rep., SR-62, 205–215, 1980.
Verbrugge RR, and Rackerd B. Evidence of talker-independent information for vowels. Lang. Sp. 1986,29,39–57.
Vickers J. The problem of induction. In: Zalta EN (Ed.), The Stanford Encyclopedia of Philosophy, 2014.
Vliegen J, and Oxenham AJ. Sequential stream segregation in the absence of spectral cues. J. Acoust. Soc. Am. 1999,105(1),339–346. doi: http://dx.doi.org/10.1121/1.424503.
von Helmholtz H. Die Lehre von Tonempfindungen. Braunschweig: Vieweg, 1862.
von Helmholtz H. On the sensations of tone as a physiological basis for the theory of music. New York: Dover, 1877.
von Humboldt W. On Language. On the Diversity of Human Language Construction and Its Influence on the Mental Development of the Human Species. Cambridge, U.K.: Cambridge University Press, 1999.

Wakita H. Estimation of vocal-tract shapes from acoustical analysis of the speech wave: the state of the art. IEEE Trans. ASSP, ASSP-27(3), 281–285, 1979.
Wang DL. On Ideal Binary Mask as the computational goal of auditory scene analysis. In: Divenyi P (Ed.), Speech Separation by Humans and Machines. New York, N.Y.: Kluwer Academic Publishers, 2004,179–196.
Wang DL, and Brown G (Eds.). Computational Auditory Scene Analysis: Principles, Algorithms and Applications. New York: Wiley-IEEE Press, 2006.
Watson CS, Wroton HW, Kelly WJ, and Benbassat CA. Factors in the discrimination of tonal patterns. I. Component frequency, temporal position, and silent intervals. J. Acoust. Soc. Am. 1975,57,1175–1185.
Watson CS, Kelly WJ, and Wroton HW. Factors in the discrimination of tonal patterns. II. Selective attention and learning under various levels of stimulus uncertainty. J. Acoust. Soc. Am. 1976,60,1176–1187.
Watson CS. Some comments on informational masking. Acta Acustica united with Acustica 2005,91(3),502–512.
Watson CS, and Kidd GR. Studies of tone sequence perception: effects of uncertainty, familiarity, and selective attention. Front. Biosci. 2007,12(1),3355–3366. http://dx.doi.org/10.2741/2318.
Wilson AS, Hall JW, and Grose JH. Detection of frequency modulation (FM) in the presence of a second FM tone. J. Acoust. Soc. Am. 1990,88(3),1333–1338. doi: http://dx.doi.org/10.1121/1.399710.
Wilson SM, Saygin AP, Sereno MI, and Ilacoboni M. Listening to speech activates motor areas involved in speech production. Nature Neurosci. 2004,7(7),701–702.
Wingfield A. Acoustic redundancy and the perception of time-compressed speech. J. Sp. Hear. Res. 1975,18(1),96–104.
Winkler R, Fuchs S, and Perrier P. The relation between differences in vocal tract geometry and articulatory control strategies in the production of French vowels: Evidence from MRI and modelling. In: Yehia HC, Demolin D, and Laboissière R (Eds.), Proc. 7th Intern. Semin. Sp. Product. Ubatuba, Brazil, 2006,509–516.
Wood S. A radiographic analysis of constriction locations for vowels. J. Phon. 1979,7,25–43.
Wundt WM. Grundzüge der physiologischen Psychologie [Foundations of physiological psychology]. Leipzig: W. Engelmann, 1874.
Yost WA, Sheft S, and Opie J. Modulation interference in detection and discrimination of amplitude modulation. J. Acoust. Soc. Am. 1989,86(6),2138–2147. doi: http://dx.doi.org/10.1121/1.398474.
Yu D, Deng L, and Acero A. Speaker-adaptive learning of resonance targets in a hidden trajectory model of speech coarticulation. Comp. Speech Lang. 2007,27,72–87.
Zurek PM. Measurement of binaural echo suppression. J. Acoust. Soc. Am. 1979,66,1750–1757.
Zwicker E. Subdivision of the audible frequency range into critical bands (Frequenzgruppen). J. Acoust. Soc. Am. 1961,33(2),248–248. doi: http://dx.doi.org/10.1121/1.1908630.

Index of terms

Author Index

Bei Fragen zur Produktsicherheit wenden Sie sich bitte an:
If you have any questions regarding product safety,
please contact:

Walter de Gruyter GmbH
Genthiner Straße 13
10785 Berlin
productsafety@degruyterbrill.com